Vitamin D in Focus

Jörg Reichrath

Vitamin D in Focus

Misconceptions Corrected

Springer

Jörg Reichrath
Klinik für Dermatologie
Universitätsklinikum des
Saarlandes
Homburg, Saarland, Germany

ISBN 978-3-662-71340-2   ISBN 978-3-662-71341-9   (eBook)
https://doi.org/10.1007/978-3-662-71341-9

This book is a translation of the original German edition "Vitamin D im Fokus " by Jörg Reichrath, published by Springer-Verlag GmbH, DE in 2024. The translation was done with the help of an artificial intelligence machine translation tool. A subsequent human revision was done primarily in terms of content, so that the book will read stylistically differently from a conventional translation. Springer Nature works continuously to further the development of tools for the production of books and on the related technologies to support the authors.

Translation from the German language edition: "Vitamin D im Fokus " by Jörg Reichrath, © Der/die Herausgeber bzw. der/die Autor(en), exklusiv lizenziert an Springer-Verlag GmbH, DE, ein Teil von Springer Nature 2024. Published by Springer Berlin Heidelberg. All Rights Reserved.

© The Editor(s) (if applicable) and The Author(s), under exclusive license to Springer-Verlag GmbH, DE, part of Springer Nature 2025

This work is subject to copyright. All rights are solely and exclusively licensed by the Publisher, whether the whole or part of the material is concerned, specifically the rights of translation, reprinting, reuse of illustrations, recitation, broadcasting, reproduction on microfilms or in any other physical way, and transmission or information storage and retrieval, electronic adaptation, computer software, or by similar or dissimilar methodology now known or hereafter developed.
The use of general descriptive names, registered names, trademarks, service marks, etc. in this publication does not imply, even in the absence of a specific statement, that such names are exempt from the relevant protective laws and regulations and therefore free for general use.
The publisher, the authors and the editors are safe to assume that the advice and information in this book are believed to be true and accurate at the date of publication. Neither the publisher nor the authors or the editors give a warranty, expressed or implied, with respect to the material contained herein or for any errors or omissions that may have been made. The publisher remains neutral with regard to jurisdictional claims in published maps and institutional affiliations.

This Springer imprint is published by the registered company Springer-Verlag GmbH, DE, part of Springer Nature.
The registered company address is: Heidelberger Platz 3, 14197 Berlin, Germany

If disposing of this product, please recycle the paper.

Preface

How important is vitamin D for our health and what do we need to do to achieve optimal supply of the "sun hormone"? Currently, hardly any other medical topic is being discussed as controversially and emotionally among scientists as well as in the public media as this question. Is vitamin D the "universal remedy from the sky pharmacy"? Or are the many reports about positive effects on our health exaggerated false reports?

Metaphorically, vitamin D deficiency can be compared to an invisible "ticking time bomb". We can neither smell, feel, nor taste it, and we also do not have any sensory organ that could recognize this danger or warn us about it. Also, because we do not feel any negative consequences at the beginning of its insidious course, the health risks associated with a vitamin D deficiency are often underestimated. Because with the increasing duration of the vitamin D deficiency, the risk for the occurrence of many different diseases increases unnoticed and unstoppable. The great importance of vitamin D for the regulation of bone and calcium metabolism is undisputed. However, this is often doubted in many other widespread diseases, including numerous infectious, autoimmune, cardiovascular, cancer, and metabolic diseases.

Vitamin D deficiency is a pandemic. According to studies by the Robert Koch Institute and other professional societies, more than 60% of the German population have a vitamin D deficiency (25-hydroxyvitamin D [25(OH)D] serum levels > 20 ng/ml)

It is convincingly shown that we do not have to choose between Scylla and Charybdis, that is, between vitamin D deficiency and increased skin cancer risk. This book provides precise recommendations for a third, healthy way and gives concrete behavioral rules for optimal vitamin D supply and healthy handling of the sun. According to current data collections, it is a tragic misdevelopment that most of us still leave the unique opportunity unused, which a sensible handling of the sun and a resulting optimal vitamin D status offer for a healthy and long life.

I hope that I have made you somewhat curious about reading this book, and wish you a lot of fun while reading!

Homburg, Germany   Jörg Reichrath

Acknowledgments

First, I would like to express my deepest gratitude to my wife Sandra and my two sons Benjamin and Niklas for their continuous support in writing this book. I would also like to extend my special thanks to Michael Holick and Armin Zittermann for providing images, and to Anna Kräz and the rest of the staff at Springer Publishing for their always trustworthy and helpful cooperation, as well as their competent support.

Contents

1 **Vitamin D and Health: Historical Considerations and Current State of Research** 1
   1.1 Misconception 1: *Vitamin D is a Vitamin* 3
       1.1.1 Correction 3
       1.1.2 Comment 3
   1.2 Misconception 2: *Because Rickets Has Largely Disappeared Today, Vitamin D Deficiency is No Longer Relevant to Health* 5
       1.2.1 Clarification 5
       1.2.2 Comment 6
   1.3 Misconception 3: *There is Scientific Evidence Refuting the Great Importance of Vitamin D for Health*................................ 30
       1.3.1 Correction 30
       1.3.2 Comment 30
   References.. 31

2 **Vitamin D Intake with Food**................... 33
   2.1 Misconception 1: *Enriching Food With Vitamin D Has No Impact on Health and Therefore Makes No Sense*........................... 33
       2.1.1 Correction 33
       2.1.2 Comment 34

	2.2	Mistake 2: *Under Our Living Conditions, the Vitamin D Requirement Is Fully Covered by the Diet—Even Without Taking Dietary Supplements*...........................	37
		2.2.1 Correction	37
		2.2.2 Comment	37
	2.3	Misconception 3: *It is a Flaw in Evolution that Breast Milk Contains Everything Infants Need for Healthy Development—With the Exception of Vitamin D*............................	40
		2.3.1 Correction	40
		2.3.2 Comment	40
	2.4	Misconception 4: *It is Necessary to take Vitamin D Together with Fatty Foods to Achieve Good Absorption in the Gastrointestinal Tract*.......	42
		2.4.1 Clarification.....................	42
		2.4.2 Comment	42
References...................................			43
3	**UV-Induced Vitamin D Production in the Skin**		45
	3.1	Misconception 1: *Vitamin D Can Be Produced in the Skin Without UV Influence*.............	46
		3.1.1 Correction	46
		3.1.2 Comment	46
	3.2	Misconception 2: *Natural Sun Exposure Allows for Sufficient Vitamin D Production in Germany in All Seasons*.................	47
		3.2.1 Correction	47
		3.2.2 Comment	48
	3.3	Misconception 3: *The Vitamin D Production of the Skin is Only Insignificantly Influenced by the Strength of the Ground-Level UV Radiation*	49
		3.3.1 Correction	49
		3.3.2 Comment	50
	3.4	Misconception 4: *Skin's Vitamin-D Synthesis is Not Influenced by Age*...................	53

		3.4.1	Correction	53
		3.4.2	Comment	53
	3.5	Misconception 5: *The Synthesis of Vitamin D Lowers the Blood Cholesterol Level by Consuming Its Precursor 7-DHC*.		53
		3.5.1	Correction	53
		3.5.2	Comment	53
	3.6	Misconception 6: *The Skin Produces Only a Few Hundred IU of Vitamin D Daily, Even Under Sufficient UV-B Influence*		54
		3.6.1	Correction	54
		3.6.2	Comment	54
	3.7	Misconception 7: *A Healthy Person Can Produce Too Much Vitamin D From Extensive Sunbathing and Thereby Endanger Health*		56
		3.7.1	Clarification	56
		3.7.2	Comment	57
	3.8	Misconception 8: *UV-B Irradiation of Dairy Cows Has No Effect on the Vitamin D Content of Cow's Milk*		58
		3.8.1	Correction	58
		3.8.2	Comment	58
	3.9	Misconception 9: *UV Irradiation of Food Has No Effect on Its Vitamin D Content*		59
		3.9.1	Correction	59
		3.9.2	Comment	59
	3.10	Misconception 10: *The Impact of Changes in Our Vitamin D Supply on Vitamin D Status is Difficult to Estimate*		60
		3.10.1	Clarification	60
		3.10.2	Comment	60
	References			62
4	**Vitamin D, Skin Types, and Sunscreen**			**63**
	4.1	Misconception 1: *The Various Human Skin Types are a Whim of Evolution Without Biological Significance—And also Without Relevance for Cutaneous Vitamin D Production*		63

		4.1.1	Correction	63
		4.1.2	Comment	64
	4.2	Misconception 2: *A Mild Sunburn is Harmless to Health*.		65
		4.2.1	Correction	65
		4.2.2	Comment	65
	4.3	Misconception 3: *The Use of Sunscreen Does Not Reduce the UV-Induced Formation of Vitamin D in the Skin*		68
		4.3.1	Correction	68
		4.3.2	Comment	69
	4.4	Misconception 4: *The Use of Sunscreens Poses No Health Risk*		70
		4.4.1	Correction	70
		4.4.2	Comment	71
	4.5	Mistake 5: *The Use of Sunscreen Does Not Pose an Environmental Burden*		79
		4.5.1	Correction	79
		4.5.2	Comment	79
	4.6	Mistake 6: *The Right Choice of Sunscreen Protects Against the Sun Without Reducing Vitamin D Synthesis*		82
		4.6.1	Correction	82
		4.6.2	Comment	82
	References			83
5	**How Does Vitamin D Work?**			**87**
	5.1	Misconception 1: *The Vitamin D Formed in the Skin or Ingested with Food is Biologically Active*		87
		5.1.1	Clarification	87
		5.1.2	Comment	88
	5.2	Misconception 2: *1,25-dihydroxyvitamin D $(1,25(OH)_2D)$ is the Only Biologically Active Vitamin D Metabolite*		88
		5.2.1	Clarification	88
		5.2.2	Comment	88

5.3		Misconception 3: *All Effects of Vitamin D Formed in the Skin Can Be Compensated by Oral Intake of Vitamin D*	91
	5.3.1	Correction	91
	5.3.2	Comment	91
References...		....	92

6 What is the Optimal Vitamin D Status? 93

6.1	Misconception 1: *To Assess Vitamin D Status, the Blood Level of the Biologically Active Metabolite 1,25-dihydroxyvitamin D $(1,25(OH)_2D)$ Should Be Determined. If this Value is Within the Normal Range, There is No Vitamin D Deficiency*	94
	6.1.1 Correction	94
	6.1.2 Comment	94
6.2	Misconception 2: *The Determination of the Serum Concentration of the Biologically Inactive Vitamin D Metabolite 25-hydroxyvitamin D (25(OH)D) is Unsuitable for Assessing a Person's Vitamin D Status*	96
	6.2.1 Clarification	96
	6.2.2 Comment	96
6.3	Misconception 3: *An Optimal Vitamin D Status is Assumed When the Blood Level for 25(OH)D is Above 10 ng/ml and When there is No Indication of a Disease of the Bone and Calcium Metabolism*	98
	6.3.1 Correction	98
	6.3.2 Comment	98
References...............................		99

7 Conclusion with Practical Recommendations for Ensuring Good Vitamin D Supply while Safely Interacting with the Sun....................... 101
References.................................. 109

Abbreviations

AhR	Aryl Hydrocarbon Receptor
DHC	Dehydrocholesterol
LXR	Liver X receptor
PPARγ	Peroxisome proliferator-activated receptor γ
ROR	Retinoic Acid Receptor-related Orphan Receptors
UV	Ultraviolet
VDR	Vitamin D receptor
Vitamin D	Calciferol
Vitamin D_3	Cholecalciferol
Vitamin D_2	Ergocalciferol
$1,25(OH)_2D_3$	1,25-Dihydroxyvitamin D_3, 1,25-Dihydroxycholecalciferol, Calcitriol
$25(OH)D_3$	25-Hydroxyvitamin D_3, 25-Hydroxycholecalciferol, Calcifediol
$20(OH)D_3$	20-Hydroxyvitamin D_3
$20,23(OH)_2D_3$	20,23-Dihydroxyvitamin D_3

Vitamin D and Health: Historical Considerations and Current State of Research

Contents

1.1 Misconception 1: *Vitamin D is a Vitamin*	3
1.2 Misconception 2: *Because Rickets Has Largely Disappeared Today, Vitamin D Deficiency is No Longer Relevant to Health*	5
1.3 Misconception 3: *There is Scientific Evidence Refuting the Great Importance of Vitamin D for Health*	30
References	31

Vitamin D deficiency is very widespread worldwide. Unfortunately, its serious dangers to our health are still significantly underestimated. This book exposes common misconceptions about the structure, function, and effect of the diverse vitamin D metabolic pathways and corrects these misconceptions according to the current state of science. The easily understandable and precise explanation of the complex relationships enables the reader to form their own opinion about the importance of vitamin D for our health. Taking into account the latest scientific findings, it is explained in detail why a single hormone can have so many different effects in almost all organs of our body.

An important reason for this lies in recently identified "new" signaling pathways that are not mediated by the classic biologically active metabolite 1,25-dihydroxyvitamin D

(1,25(OH)$_2$D), but by other active vitamin D metabolites (including 20,23(OH)$_2$D$_3$). These largely unexplored metabolites bind and regulate not only the classic vitamin D receptor (VDR), but also numerous other corresponding receptor molecules, including aryl hydrocarbon receptor (AHR), liver X receptor (LXR) and peroxisome proliferator-activated receptor (PPAR). These nuclear receptors regulate as transcription factors more than 1000 genes of our genome. As a result, they perform many important functions in our body that have only been associated with vitamin D in recent years. Since these receptors are located in almost all tissues of our body, biologically active vitamin D metabolites can exert their physiological effects in so many different organs of our body.

In addition, previously overlooked functions of vitamin D such as the regulation of our internal clock and the defense against environmental toxins are explained. It is explained in detail what is meant by optimal vitamin D supply or an optimal vitamin D status. The reader learns why the different vitamin D signaling pathways have different threshold values for a sufficient/optimal vitamin D status. Thus, rickets and other diseases of bone and calcium metabolism only occur at very low vitamin D status (25(OH)D serum concentration <10 ng/ml). However, at 25(OH)D serum levels of 10–20 ng/ml, there is already an increased risk of occurrence and unfavorable clinical course of numerous infectious, cancer, autoimmune and metabolic diseases.

Since only a few foods contain significant amounts of vitamin D, about 80–95% of the required vitamin D must be formed in the skin with the help of UV-B radiation. This book also answers the central question: How much sun is good for us? It provides individual recommendations for optimal vitamin D supply through healthy sun exposure and—if necessary—through the supplementary intake of vitamin D-containing dietary supplements. In addition, there are often surprising answers to questions such as: *Why does breast milk contain everything a baby needs for life, but almost no vitamin D? Was vitamin D the driving force in the development of different skin colors? What does sunscreen have to do with coral bleaching?*

1.1 Misconception 1: *Vitamin D is a Vitamin*

1.1.1 Correction

According to the classic definition, vitamins are substances that must be supplied to our body from the outside. Since vitamin D can be formed in sufficient quantities in our skin from precursors (the cholesterol precursor 7-dehydrocholesterol, 7-DHC), this requirement is not met. Therefore, strictly speaking, the designation of vitamin D as a vitamin is not correct (Reichrath 2021; Wacker and Holick 2013).

1.1.2 Comment

1.1.2.1 Historically, How did the Misnomer of Vitamin D as a "Vitamin" Come About?

In the early decades of the 20th century, British physician Edward Mellanby (1884–1955) and chemist Elmer Verner McCollum were able to artificially induce rickets in dogs and rats through malnutrition with specially prepared food, and then cure it in all cases by administering cod liver oil. This led to the assumption that cod liver oil must contain an *"anti-rachitic vitamin"*. Mellanby initially believed the recently discovered vitamin A in cod liver oil to be the triggering factor. However, since cod liver oil was still able to cure rickets even after oxidative treatment, which destroys vitamin A, it quickly became clear that a vitamin A deficiency could not be a possible cause of rickets. McCollum therefore postulated that another substance independent of vitamin A must be responsible for this effect. The then still unknown substance was named *"Vitamin D"*, as the presumed fourth identified vitamin (after vitamins A, B, and C) (Reichrath 2021; Wacker and Holick 2013).

1.1.2.2 What Class of Substance is Vitamin D?

Biochemically, vitamin D (cholecalciferol, calciol) is the less effective fat-soluble precursor (prohormone) of the biologically

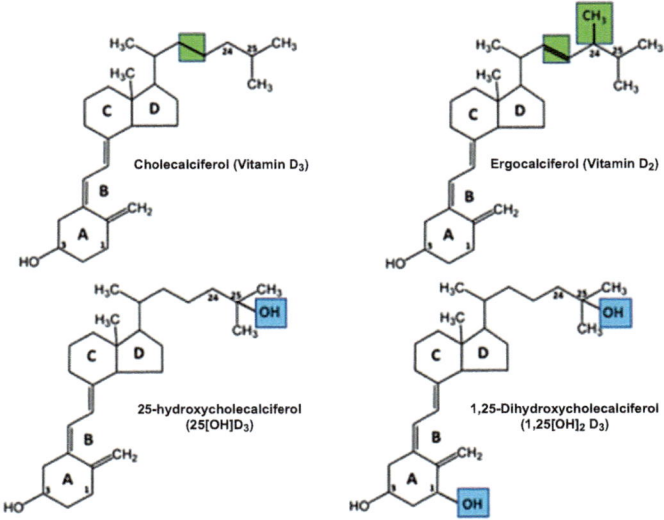

Fig. 1.1 Chemical structural formulas of vitamin D and important metabolites. Green rectangle: areas that differ between vitamin D_2 and D_3. Blue rectangle: important functional groups. (*Source*: Zittermann 2022b)

active steroid hormone 1,25-dihydroxyvitamin D. The basic structure of steroid hormones, which also include cortisone and sex hormones like estrogen, consists of four combined ring structures, referred to as *ring A* to *ring D* (Fig. 1.1). Vitamin D belongs to the group of so-called seco-steroid hormones. These are formed by a process called photolysis, in which one of these rings (usually the *B-ring*) is opened by high-energy ultraviolet (UV) radiation.

Vitamin D is not a precisely defined single chemical substance, but the umbrella term for five different forms:

Vitamin D_1: (formed from ergocalciferol [vitamin D_2] and lumisterol)
Vitamin D_2: (calciferol or ergocalciferol)
Vitamin D_3: (cholecalciferol, colecalciferol or calciol)

Vitamin D_4: (22,23-dihydroergocalciferol, the saturated form of vitamin D_2)
Vitamin D_5: (sitocalciferol)

The two most important representatives are vitamin D_3 and vitamin D_2 (Fig. 1.1). While vitamin D_3 is formed in the skin or taken in through animal foods, vitamin D_2 enters the body through plant products. When vitamin D is mentioned, it usually refers to vitamin D_3. This is the natural vitamin D that is produced in the human body. The formation of biologically active metabolites requires subsequent metabolic steps, which apply equally to vitamin D_2 and D_3. The starting point of vitamin D biosynthesis is 7-dehydrocholesterol(7-DHC), which is stored in large quantities in the membranes that surround the cells in the skin. For a good supply of vitamin D, sufficient exposure of the skin to UV-B radiation is crucial (Reichrath 2021; Wacker and Holick 2013).

► **Note**
Vitamin D is not a vitamin, as it can be produced in sufficient quantities in the body and does not necessarily have to be supplied from outside.

1.2 Misconception 2: *Because Rickets Has Largely Disappeared Today, Vitamin D Deficiency is No Longer Relevant to Health*

1.2.1 Clarification

Even in today's times, vitamin D deficiency increases the risk of both diseases related to bone and calcium metabolism such as rickets and many other diseases, including numerous cancers, cardiovascular, infectious, or autoimmune diseases. Moreover, it also affects their unfavorable course (Reichrath 2021; Wacker and Holick 2013).

1.2.2 Comment

1.2.2.1 Sun, Vitamin D, Rickets and Cancer: A Historical Perspective

Our current knowledge about the diverse positive effects of sun and vitamin D on our health has been significantly shaped by groundbreaking scientific findings in two completely independent disease areas (Reichrath 2021; Wacker and Holick 2013). These new insights, which were pioneering at the time, initially concerned the bone disease rickets and centuries later also cancer diseases. In both cases, striking parallels can be found: In the research of rickets as well as cancer diseases, hypotheses were initially established based on careful clinical observations, which were initially criticized and rejected by science, but later impressively confirmed (Reichrath 2021; Wacker and Holick 2013).

Rickets (from Greek ῥάχις *rháchis,* back, spine; synonym: English disease, English *Rickets*), also referred to as *Rhachitis* in older texts—etymologically correct—is a disease of bone growth occurring in children, characterized by disorganization of the growth plates and disturbed mineralization of the bones (Fig. 1.2). Archaeological examinations of Egyptian mummies and historical bone finds unequivocally prove that rickets has existed in human evolution from early prehistory at all times and in all parts of the world.

> **Note**
> Vitamin D deficiency is associated with an unfavorable course of both diseases related to bone and calcium metabolism (such as rickets) and many cancers, cardiovascular, infectious, or autoimmune diseases, as well as with an increased risk of their occurrence.

1.2.2.2 Description of Rickets as a Distinct Disease by Hieronymus Reusner, Daniel Whistler, and Francis Glisson

Rickets was first described as a distinct disease in the 16th century by Hieronymus Reusner (see Reichrath 2021; Wacker and

1.2 Misconception 2: *Because Rickets ...*

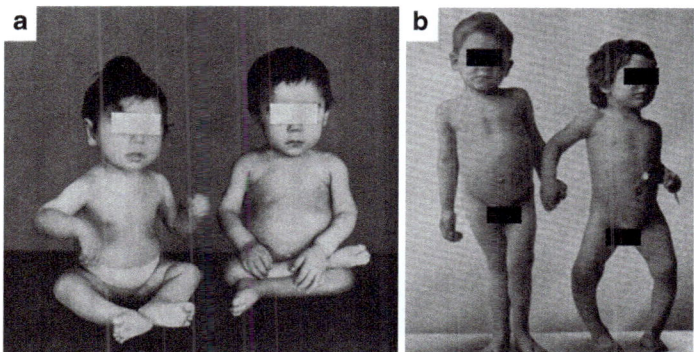

Fig. 1.2 Bone deformities in the classic vitamin D deficiency disease rickets. (**a**) 10 months (girl, left) and 2.5 years (boy, right) old siblings from the 1930s. (**b**) The same siblings 4 years later with the X- and O-legs typical for rickets. Note the severe bone deformities in the area of the lower legs. (*Source*: Wacker and Holick 2013, with kind permission from Dr. Holick)

Holick 2013). Even then, it was known that this disease could be successfully treated with cod liver oil, which is particularly rich in vitamin D. The connection between vitamin D deficiency and rickets has been more closely researched since the mid-17th century. Around 1620, Daniel Whistler documented the occurrence of this disease in the English counties of Somerset and Dorset (see Reichrath 2021; Wacker and Holick 2013). He found that children growing up in industrial cities in Great Britain often had a small stature and numerous bone deformities, particularly in the area of their lower legs (Fig. 1.2). He referred to this disease as *Morbus Anglorum* ("English disease"), a term that was used until the 19th century. At that time, smog caused by private (wood burning) and industrial environmental pollution regularly settled over cities in valley and basin locations (Fig. 1.3). This partially shielded the skin from the UV radiation of the sun necessary for vitamin D synthesis. As a result, many children showed the typical symptoms of rickets, which also affected the English upper class. In 1650, Francis Glisson described the disease in detail and accurately for the first time (see Reichrath 2021; Wacker and Holick 2013).

Fig. 1.3 Dimly lit alley of the Scottish industrial city of Glasgow around 1870, as was typical in the 19th and early 20th centuries. (*Source*: Wacker and Holick 2013)

1.2.2.3 Sniadecki's Hypothesis: Rickets as a Disease Caused by Too Little Sun

Rickets continued to plague people in the industrial cities of Europe and North America for another long 200 years. It was not until 1822 that the Polish doctor Sniadecki hypothesized that a lack of sunlight was the reason why his young patients, whom he cared for in the big city of Warsaw, were significantly more likely to suffer from rickets than those he cared for in rural regions (see Reichrath 2021; Wacker and Holick 2013).

1.2.2.4 Curing Rickets with Natural and Artificial UV Radiation by Raczyński, Huldschinsky, Unger, and Hess

However, it took almost another 100 years for these careful observations to be appropriately recognized. In 1912, the Polish pediatrician and researcher Jan Rudolf Raczyński sent children with rickets to the Carpathians for treatment. He had also postulated a connection between the development of rickets and a lack of sunlight (see Reichrath 2021; Wacker and Holick 2013). In 1889, Theobald Palm published his observation that children growing up in the dirtiest living conditions in India and other parts of Asia rarely suffered from rickets. In contrast, this bone disease was very common in the industrial cities of Great Britain and the USA. Estimates suggest that at the turn of the 20th century, up to 90% of children in North American industrial cities like Boston and New York suffered from rickets and the long-term consequences of the associated bone deformities (see Reichrath 2021; Wacker and Holick 2013).

In 1918/19, the German pediatrician Kurt Huldschinsky completely cured young rickets patients by UV irradiation with a mercury arc lamp (see Reichrath 2021; Wacker and Holick 2013). Shortly thereafter, in 1921, Alfred Fabian Hess (1875–1933) and L. F. Unger reported on the successful treatment of children suffering from rickets through sunbathing on the roof of a New York hospital (see Reichrath 2021; Wacker and Holick 2013). By around 1930, it was generally recognized in Europe and the USA that children could be effectively protected from rickets through moderate sunbathing.

Americans Harry Steenbock, Alfred Fabian Hess, and L. F. Unger subsequently found that both the skin of mammals and various plant and other foods contain a substance (provitamin) that becomes the "anti-rachitic vitamin" when irradiated. Consequently, not only humans but also foods, including milk for children, have since been treated with UV rays. The regular administration of vitamin D-rich cod liver oil or vitamin D as a dietary supplement in children has become common practice worldwide. It has largely eliminated the formerly widespread rickets, the best-known vitamin D deficiency disease, in many regions. However, a large part of the population, especially older people, still have low vitamin D levels. These are associated with both the increased occurrence and the unfavorable course of numerous cancer and other diseases. According to the current assessment of many experts, the current data situation impressively demonstrates the enormous potential that improving vitamin D supply has for promoting the health of our population.

1.2.2.5 Does Sun Protect Against Cancer? Initially Just a Hypothesis

The hypothesis that sun protects against cancer was initially based on various individual clinical and epidemiological observations. Only later were the causes systematically researched and a connection between the effect of the sun's rays and vitamin D established. The German-American statistician Frederick Ludwig Hoffman (born 1865 in Varel/Germany, died 1946 in San Diego/USA) showed a possible connection (association) between the effect of the sun and the risk of dying from cancer (mortality) for the first time in 1915. He examined the number of cancer deaths worldwide in cities located at different latitudes and showed a significantly lower number of deaths in cities with a sunny geographical location (see Reichrath 2021; Wacker and Holick 2013). The numbers published by Hoffman documented a clear decrease in cancer deaths with increasing proximity of the place of residence to the equator (residence latitude 60° North–50° North: 105.7 cancer deaths/100,000 inhabitants; latitude 10° North–10° South: 40.9 cancer deaths/100,000 inhabitants). Today we know that with increasing proximity to the

1.2 Misconception 2: *Because Rickets …*

equator, both the strength of the ground-level UV radiation and the vitamin D status of the people living there increase.

Peller and Stephenson then made another interesting observation in the 1930s (see Reichrath 2021; Wacker and Holick 2013). They found in sailors of the US Navy a striking discrepancy between the frequent occurrence of non-melanoma skin cancer, often caused by the cumulative effect of sunlight over many years, and the rare occurrence of much more dangerous other cancers. Peller and Stephenson observed that the rate of non-melanoma skin cancer (basal cell carcinoma [Fig. 1.4] and squamous cell carcinoma [Fig. 1.5]) in sailors of the US Navy who worked on deck and were heavily exposed to the sun was about 8 times higher than that of sailors who were predominantly below deck. However, the total number of those who died of other types of cancer was about 60% lower among the sailors working on deck than in the general population, which served as a control group. After further investigations, Peller and Stephenson published in 1937 their observation that people with non-melanoma skin cancer were significantly less likely to suffer from other types of cancer than comparison groups of people who did not suffer from non-melanoma skin cancer. From these observations, Peller concluded that the sun's rays, which cause skin cancer, could simultaneously confer a certain immunity against other cancers of internal organs. Since non-melanoma skin cancer is generally not life-threatening and curable, he recommended inducing it to protect oneself from other, far more dangerous cancers (see Reichrath 2021; Wacker and Holick 2013).

The observations of Peller and Stephenson were picked up just a few years later by the American pathologist Frank Apperly. He published in 1941 in North America an inverse relationship between the intensity of ground-level UV radiation and mortality rates from cancer (see Reichrath 2021; Wacker and Holick 2013). Apperly reported that Americans living in relatively sun-poor states of relatively high geographical latitudes, including New Hampshire, Vermont, and Massachusetts, had a significantly higher risk of dying from cancer compared to people living in sunnier southern states like Alabama and Georgia. In the sun-rich areas of relatively low latitudes near the equator, cancer

Fig. 1.4 Basal cell carcinoma (basalioma) of the skin. (**a**) Nodular variant. (**b**) Superficial variant. (*Source*: Plewig 2016)

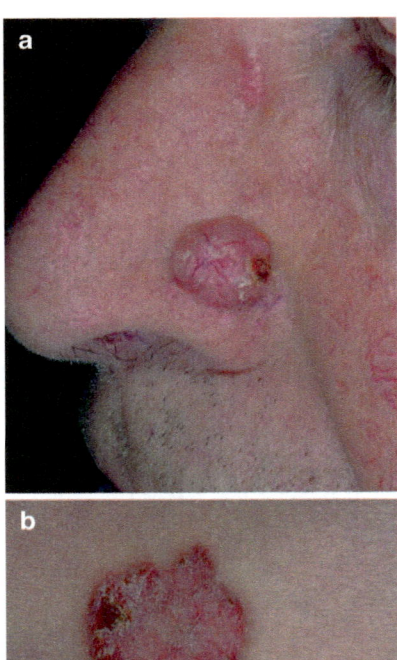

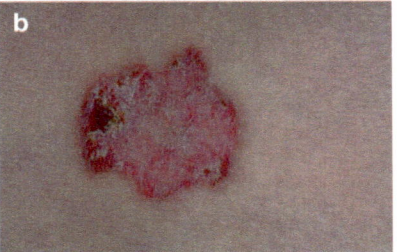

mortality rates were the lowest. The relationship between the geographical latitude of the place of residence and cancer mortality led Apperly to assume that sunlight could have a directly positive effect on the immune status. However, the publications with these observations and hypotheses were largely ignored by large parts of medicine and science at first.

▶ **Note**

In the research of the bone disease rickets and centuries later of cancers, there are striking parallels to the importance of sun and vitamin D: Both in rickets and in cancers, hypotheses were initially set up based on careful clinical

1.2 Misconception 2: Because Rickets ...

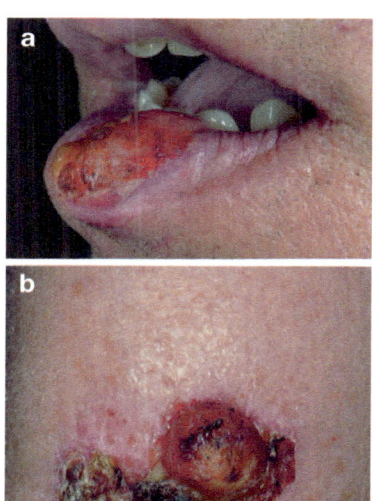

Fig. 1.5 Squamous cell carcinoma of the skin. (**a**) On the lower lip. (**b**) On light-damaged skin (*Source*: Plewig 2016)

observations, which were initially criticized and rejected by science, but later impressively confirmed.

1.2.2.6 The Garland Brothers, Edward D. Gorham, and the Vitamin D-Cancer Hypothesis

The American epidemiologists Cedric F. Garland, Frank C. Garland, and Edward D. Gorham first proposed what is often referred to as the "Vitamin D-Cancer Hypothesis" in the early eighties of the last century. This hypothesis first showed a connection between Vitamin D and the cancer-protective effect of sunlight. They reported that people living in regions with little sunlight, and therefore have a low Vitamin D status, have an increased risk of developing several dangerous types of cancer. These include, among others, colorectal, breast, and ovarian cancer, and interestingly, also malignant melanoma (see Reichrath 2021; Wacker and Holick 2013). Later, an effect of Vitamin D against cancer cells was found in prostate cancer and many other tumor diseases (see Reichrath 2021; Wacker and Holick 2013). Vitamin D deficiency is widespread worldwide and is particularly common in cancer patients. For example, a study diagnosed a Vitamin D_3 deficiency in 59% of colorectal cancer patients, which was also associated with an unfavorable prognosis.

> **Note**
> The American epidemiologists Cedric F. Garland, Frank C. Garland, and Edward D. Gorham proposed in the early 1980s what is often referred to as the "Vitamin D-Cancer Hypothesis". This postulates that Vitamin D deficiency, as a result of low UV-B radiation in people living at high latitudes with little sunlight, is responsible for the high mortality from colorectal, breast, and ovarian cancer shown there.

1.2.2.7 Newer Research Supports the Vitamin D-Cancer Hypothesis

These studies were later picked up and supplemented by William Grant. He reported that the number of deaths from several types of cancer could be significantly reduced by adequate exposure of the population to sunlight. Grant estimated that over a period of 24 years (1970–1994), a total of 566,400 US Americans died prematurely from 13 types of cancer he studied due to insufficient UV exposure. These results are supported by newer studies. More recent estimates suggest that due to inadequate UV exposure and vitamin D deficiency, between 50,000 and 63,000 US Americans and 19,000–25,000 Britons die prematurely from cancer each year. In recent years, a multitude of experimental laboratory results and other studies convincingly support the "Vitamin D-Cancer Hypothesis" founded by Garland et al., demonstrating that adequate vitamin D supply is important for

1.2 Misconception 2: Because Rickets ...

protection against cancer (see Reichrath 2021; Wacker and Holick 2013; Kuznia et al. 2023; Niedermaier et al. 2022).

A systematic literature review conducted a few years ago at the German Cancer Research Center (DKFZ) in Heidelberg yielded surprising results (Kuznia et al. 2023; Niedermaier et al. 2022). The results of this study suggest that by taking vitamin D, cancer mortality in the German population could be reduced by 12%—provided the vitamin is taken daily. The researchers examined the entire scientific literature in a so-called meta-analysis. They identified a total of 14 highest quality studies, whose approximately 105,000 participants were randomly assigned to groups that regularly took either vitamin D_3 or placebo. Although the summary of all 14 studies showed no statistically significant results, when the studies were divided according to whether the vitamin D_3 intake was daily in relatively low dosage (400 to 4000 IU per day) or as a less frequently administered high single dose (60 000 to 120,000 IU once per month or less frequently), a large difference was observed (Kuznia et al. 2023; Niedermaier et al. 2022). In the four studies with the high single doses, no effect on cancer mortality was observed. In the summary of the ten studies with daily dosing, however, the researchers found a statistically significant reduction in cancer mortality by 12% (Kuznia et al 2023; Niedermaier et al. 2022). This 12% reduction in cancer mortality was observed after indiscriminate vitamin D_3 administration to individuals with and without vitamin D deficiency. Therefore, the authors of the study assume that the cancer-protective effect is considerably higher for those people who actually have a vitamin D deficiency (Kuznia et al. 2023; Niedermaier et al. 2022). The better efficacy of the daily vitamin D_3 doses can, in the authors' view, be explained by the more regular bioavailability of the active ingredient, the hormone 1,25-dihydroxyvitamin D. Because 1,25-dihydroxyvitamin D, which can likely inhibit tumor development and growth through various mechanisms, is only produced in the body by conversion of its precursor vitamin D.

A more detailed analysis of the studies with daily intake also revealed that people from the age of 70 benefit most from vitamin D intake. In addition, the effect was most pronounced when vitamin D intake began before the cancer diagnosis. In the authors' view, this work underscores the great potential of vitamin D administration in the prevention of cancer deaths. They emphasized that regular intake in low doses is associated with almost negligible risk and very low costs.

▶ **Note**
Newer research suggests that daily vitamin D intake could reduce cancer mortality in the German population by 12%.

1.2.2.8 Does Vitamin D Deficiency Favor Other Diseases Besides Cancer?

In recent years, an increasing number of research reports convincingly demonstrate an important role of vitamin D deficiency not only in cancer, but also in many other diseases. This particularly applies to health disorders for which a positive effect of sunlight has previously been shown. The UV-induced vitamin D production in the skin is now attributed with so many positive effects on our health that the description of the sun as a "universal remedy from the heavenly pharmacy", formulated as early as the 19th century by the German writer August von Kotzebue, seems to be justified. According to estimates from a published study, about 12% of all deaths in the USA (340,000 people per year) can be linked to insufficient sunlight. One of the possible causes for this correlation is primarily the vitamin D deficiency caused by too little sun.

1.2.2.8.1 Avoiding the Sun May Pose a Risk Factor for Overall Mortality on the Same Scale as Smoking

Interesting results were provided in 2011 by an observational study originally conducted to investigate malignant melanoma (skin cancer) in relatively sun-poor regions of southern Sweden over a period of 20 years involving 29,518 women ("*Melanoma in South Sweden*" [MISS] Study). In this study, the women were divided into 4 groups: from "*sun avoiders*" to "*sun worshippers*". Surprisingly, the risk of dying was significantly increased among the "*sun avoiders*" (compared to the "*sun worshippers*").

1.2 Misconception 2: *Because Rickets ...*

These differences were dose-dependent and were on a scale similar to that seen in smokers compared to non-smokers. Thus, mortality was twice as high in women who avoided sunlight compared to the "sun worshippers". Women who had spent at least one "sun and beach holiday" per year over three decades had a 30% lower overall mortality rate. The mortality risk of non-smoking women who avoid the sun was comparable to that of smoking women whose lifestyle was associated with the highest sun exposure. The authors interpreted their results to mean that avoiding sunlight poses a risk factor for overall mortality on the same scale as smoking.

The same research group then showed in a subsequent examination of the women included in this study that participants with light skin type (suggesting better vitamin D supply) see also Chap. 3 had an 8% lower mortality rate than women with darker skin type.

▶ **Note**
Study results suggest that consistently avoiding sunlight poses a risk factor for overall mortality on the same scale as smoking.

1.2.2.8.2 Vitamin D Deficiency as a Risk Factor for Cardiovascular Diseases

In a further follow-up examination of this study cohort, it was found that the shorter life expectancy of the participating women was largely caused by deaths from cardiovascular diseases. An American observational study (Framingham Heart Study) had previously clarified that vitamin D deficiency is an important risk factor for cardiovascular diseases. The importance of vitamin D for heart health has since been confirmed by numerous studies. This includes a recent study that evaluated data from over 260,000 people from the UK Biobank (Ang et al. 2021). The UK Biobank is a large-scale sample and data collection of the British population. Its aim is to promote health-related research. According to the results of this study, people with vitamin D deficiency (25(OH)D concentration in the blood of <25 nmol/L [10 ng/ml]) have an 11% higher risk of dying from cardiovascular diseases than people with sufficient vitamin D status (25(OH)D concentration in the blood of >50 nmol/L. [20 ng/ml]). Depending on the vitamin D level, blood pressure also changed in a comparable pattern. However, the positive effect of vitamin D on cardiovascular health was only noticeable up to a 25(OH)D threshold value of 50 nmol/L (20 ng/ml). From a level of 50 nmol/L (20 ng/ml), the risk of cardiovascular diseases did not decrease significantly with a further increase. The authors of the study concluded that vitamin D improves blood circulation and lowers blood pressure, thereby significantly contributing to a reduced risk of cardiovascular diseases.

1.2.2.8.3 Vitamin D Deficiency and Immune System

Our body's immune system is responsible for defending against unwanted intruders such as bacteria, fungi, and viruses. It is excellently equipped for this. Figuratively speaking, it has an efficient, well-equipped, sophisticated system of watch and control posts and reporting points, as well as differently armed armies. The immune system can be described as a complex network of different cell types, which are located in various organs and perform various precisely coordinated functions. The skin, as the body's border organ to the outside, is an important defense organ against various threats lurking in our environment and an important part of this immune system. It contains many of the cell types involved in the immune response (including Langerhans cells, so-called antigen-presenting cells, which are mainly involved in the development of allergies; leukocytes, T and B lymphocytes, macrophages and other inflammatory cells) and maintains a close relationship with other important components of the immune system, such as the lymph nodes.

The immune system also uses the so-called HLA system, a recognition system present on all cells, which can be compared to a fingerprint or identity card. This recognition system ultimately determines whether the immune system fights a substance (often proteins) or cell (immunity) or leaves it alone (tolerance). This particularly important task is performed by the

1.2 Misconception 2: *Because Rickets ...*

immune system continuously and with every contact with endogenous or exogenous substances. The decision between immunity (defense reaction, often accompanied by inflammatory reactions) and tolerance (no defense) is of great importance for the development of allergies and autoimmune diseases. The cell types involved in these processes not only have local effects. They can also migrate to other organs or release messenger substances (e.g., antibodies) that are transported via the blood and thus trigger immune responses at other, distant locations. Numerous different cell types ultimately carry out the immune system's defense reaction (immunity).

It is certainly understandable that a system as complex as this involves a multitude of different cell types and that its smooth functioning depends on the error-free execution of decision-making processes at numerous important interfaces. Sunlight and the vitamin D system also intervene in the immune response at many of these important interfaces. Given the great importance of the skin as an immune organ, it is not surprising that sun exposure of the skin also significantly influences our immune system. The sunlight exerts many different effects on our immune system. Depending on many factors (including the dose and spectrum of sunlight—e.g., the proportion of UV-A, UV-B; the age of the person exposed to sunlight; the condition of the skin), the sun regulates the immune system in different directions, it can strengthen or weaken it. Therefore, I would like to emphasize at this point that assessing the health effects of sunlight on our immune system is not simple. These depend very much on its dose and other factors. A general evaluation of the effect of sunlight or vitamin D metabolites as "*immune system inhibiting*" (favoring tolerance) or as "*immune system strengthening*" (favoring immunity) is not possible (Fig. 1.6). Likewise, a blanket evaluation of the "*inhibitory*" effects as undesirable or even harmful and the "*strengthening*" effects as desirable and healthy is not possible. I would like to illustrate this with the following example: The "*inhibitory effect*" of UV radiation on inflammatory cells can be desirable in certain situations, as it can counteract the development of allergies. In other situations, such as in infection defense, the same "*inhibitory effect*" of UV

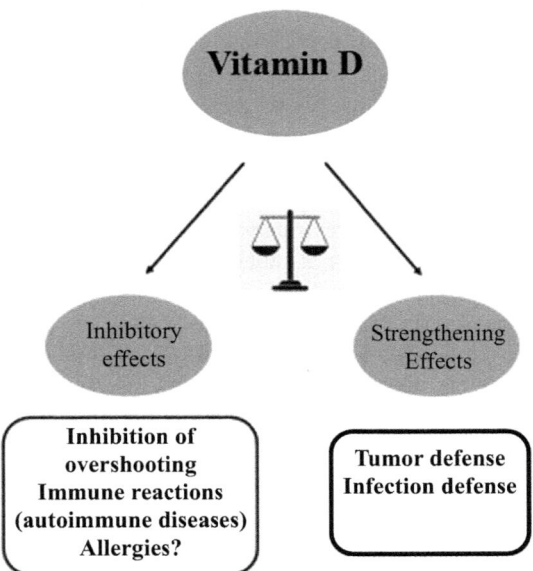

Fig. 1.6 Vitamin D: a "guardian of the immune system". Note: Depending on the need, vitamin D can both weaken and strengthen immune reactions

radiation on inflammatory cells is undesirable, as it can be associated with a higher risk of infectious diseases or their unfavorable courses.

1.2.2.8.4 The Groundbreaking Studies of Margret Kripke's Research Group

A crucial advancement for understanding the effect of the sun on our immune system was provided in the 1970s by the groundbreaking studies of Margret Kripke's research group (see Reichrath 2021). She was able to show that irradiation of the skin with UV-B in mice can prevent the development of a so-called contact allergy (a typical example of a contact allergy is fashion jewelry intolerance in nickel allergy sufferers). Later, Margret Kripke and other research groups were able to show

1.2 Misconception 2: Because Rickets ...

that both the wavelength of the UV radiation and its dose are of great importance for this influence on the development of allergies. Daily irradiation of mice with low UV-B doses ("low-dose model", about 100–200 mJ/cm^2) already caused a so-called immune tolerance after several days. Thus, the mice irradiated in this way developed an allergy less frequently after applying an allergenic substance (allergen) to the skin compared to the non-irradiated mice in the control group.

If mice are irradiated in a so-called half-side experiment with UV-B doses of 500–3000 mJ/cm^2 ("high-dose model"), this also causes a so-called immune tolerance in the non-irradiated skin. This phenomenon clearly contradicts the assumption that UV-B-mediated immune tolerance is exclusively mediated by local effects in the directly irradiated skin. Probably certain messenger substances or cells are released in the irradiated skin. These then circulate via the blood into non-irradiated skin areas and thereby also cause immune tolerance against allergens there.

These research results are supported by many further clinical studies (see Reichrath 2021). In one of these studies, voluntary subjects were divided into two groups and initially either irradiated with 144 mJ/cm^2 UV-B for 4 consecutive days or not irradiated (control group). Immediately afterwards, an attempt was made to cause a contact allergy in the skin by applying a potent allergen (in this case dinitrochlorobenzene). Four weeks later, about 60% of the UV-B-irradiated subjects showed an allergy. In the control group, on the other hand, a contact allergy was found in all study participants. Other research groups were also able to show a reduction in the rate of contact allergies and an induction of tolerance in similar studies after UV-B irradiation. Overall, the statement "the dose makes the poison" applies to the influence of sunlight on the development of allergies. It should also be emphasized once again that the entire evolution of humans took place under the influence of the sun. Therefore, it is plausible that a "natural" handling of the sun is of great importance for the developing immune system in children as well as later for the error-free function of the mature immune system in adults. A lack of sun could be one reason why both adults and children are increasingly suffering from allergies.

1.2.2.8.5 Vitamin D Deficiency and Multiple Sclerosis (MS)

Many research results suggest the importance of vitamin D deficiency as a risk factor for autoimmune diseases such as Multiple Sclerosis (MS). Already in 2006, a study found that a low vitamin D level is associated with an increased risk of MS (Munger et al. 2006). Other studies also suggest the importance of vitamin D status for the clinical course of MS. Recent studies show that the daily intake of about 7000 IU of vitamin D can reduce the occurrence of characteristic spinal cord damage (demyelination lesions).

1.2.2.8.6 Vitamin D Deficiency in Infectious Diseases

Numerous studies suggest an important role of vitamin D in the defense against infectious diseases (Martineau et al. 2017). Much-noticed research results of recent years show that UV-induced vitamin D production in the skin supports and strengthens important functions of the immune system. This also includes the defense against pathogens and the destruction of cancer cells. These effects are supported by the UV-B-induced formation of vitamin D in the skin (see Reichrath 2021). Thus, the "sun hormone" increases the formation of proteins in various cell types of the immune system (e.g., macrophages), which are referred to as defensins or antimicrobial peptides. These antimicrobial peptides can not only kill bacteria, but probably also sustainably strengthen the immunological defense against cancer cells. The formation of antimicrobial peptides in macrophages and other cell types of the immune system caused by vitamin D is also partly responsible for Nils Riedberg Finsen (1860–1904) being able to successfully treat skin tuberculosis with sun rays. In 1903, this Danish doctor received the Nobel Prize in Medicine for the treatment of skin tuberculosis (lupus vulgaris) with sun rays focused through optical lenses (see Reichrath 2021).

In a large meta-analysis (25 studies, data from a total of 10,933 study participants from 14 countries on four continents, age up to 95 years), it was shown that the daily or weekly administration of vitamin D can prevent acute respiratory infections (Martineau et al. 2017). The protective effect of this supplementation was particularly pronounced at 25(OH)D starting concentrations in the blood of less than 25 nmol/L (10 ng/ml). However, a protective effect of vitamin D administration could also be detected at 25(OH)D concentrations of more than 25 nmol/L. Overall, according to the study results, the intake of vitamin D protected one in 33 participants from acute respiratory tract infections. The authors suspected that vitamin D promotes the formation of antimicrobial peptides and primarily strengthens the innate immune system. Since acute respiratory infections, according to the authors, caused 2.65 million deaths worldwide in 2013 and were also responsible for 10% of outpatient or inpatient doctor visits in the United States, the authors advocated considering the enrichment of certain foods with vitamin D. This is particularly important in regions where vitamin D deficiency is widespread.

> **Note**
> In 1903, the Danish doctor Nils Riedberg Finsen (1860–1904) received the Nobel Prize in Medicine for the treatment of skin tuberculosis (lupus vulgaris) with sun rays focused through optical lenses.

1.2.2.9 The UV-induced Synthesis of Vitamin D Metabolites in the Skin and Subsequent Vitamin D Signaling Pathways

For good vitamin D supply, sufficient UV exposure of the skin is crucial. The starting point of cutaneous vitamin D biosynthesis is the sterol 7-dehydrocholesterol (7-DHC) (Fig. 1.7). This cholesterol precursor is produced in large quantities in the skin (about 2000 ng/cm^2) and then stored in the outer membranes of epithelial cells (keratinocytes) of the epidermis. Through a photochemical reaction known as photolysis, 7-DHC in the membranes of the cells of the upper layer of the skin (epidermis) is first converted into the thermodynamically unstable previtamin D$_3$, where the B-ring is broken under the influence of UV-B radiation (280–320 nm). Through a subsequent time- and temperature-dependent chemical reaction (so-called *sigmatropic shift* of a proton from C-19 to C-9 with subsequent isomerization),

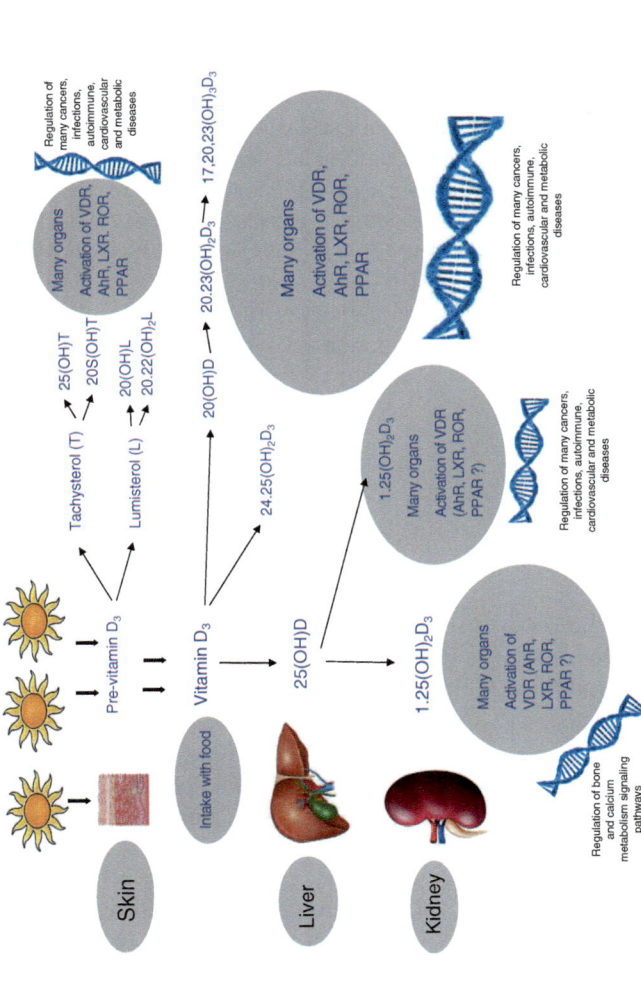

Fig. 1.7 Simplified overview of the UV-induced synthesis of vitamin D metabolites in the skin and subsequent vitamin D signaling pathways

previtamin D becomes V*itamin D_3*. The action spectrum of UV rays in the conversion of 7-DHC to previtamin D peaks at about 298 nm. If a certain amount of 7-DHC is exposed to simulated equatorial sunlight in a test tube experiment, about 20% of the starting amount is converted to previtamin D_3 within a few minutes. This amount of previtamin D_3 remains in equilibrium with continued UV-B irradiation, as *previtamin D_3* is also photolabile and is rapidly degraded by optical radiation. More precisely, with continued UV-B irradiation, previtamin D_3 is degraded to *lumisterol* and *tachysterol*, (metabolic products, within eight hours before it can isomerize to vitamin D_3,), which were until recently thought to be largely physiologically inactive. Thus, even with strong sun exposure, an excessive accumulation (cumulation) of vitamin D_3 in the skin is prevented by the subsequent formation of these by-products. The vitamin D_3 formed from previtamin D_3 is also rapidly degraded under the influence of optical radiation and is therefore photolabile. If vitamin D_3 is not quickly transported out of the skin via the bloodstream, at least three further metabolites are formed from it under the continued influence of UV-B and UV-A radiation (up to 345 nm), which are currently believed to be biologically less active: *suprasterol-1* and *-2* and *5,6-transvitamin D_3*. Therefore, even the most intense sun exposure cannot cause vitamin D intoxication (poisoning) in healthy people. Thus, a short sun exposure (with a sufficiently high UV-B component) for a few minutes produces a comparable amount of vitamin D_3 as exposure over a longer period. This protects the body from vitamin D intoxication due to excessive UV radiation.

The (classical) endocrine $1,25(OH)_2$ D_3/Vitamin D receptor (VDR) pathway

Only through downstream chemical reactions taking place in numerous organs (mediated by several so-called P450 enzymes, i.e., the addition of hydroxyl groups [OH]), especially in the liver (via CYP27A1 and CYP2R1) and kidney (via CYP27B1), is vitamin D converted into 1,25-dihydroxyvitamin D ($1,25(OH)_2D_3$, calcitriol), the classical, biologically active vitamin D hormone, as well as other biologically active metabolites. For this purpose, the vitamin D_3 formed in the skin or ingested with food first travels via the bloodstream to the liver, bound to vitamin D binding proteins (DBP, GC), where it is converted into 25-hydroxyvitamin D_3 (25(OH)D_3). This metabolic step has a high capacity, is not limited and is little regulated. This means that the vitamin D circulating in the blood is completely converted into 25(OH)D. 25(OH)D_3 is the biologically less active main circulation form of the vitamin D metabolites in the blood and the most important laboratory parameter for assessing vitamin D status. The determination of the concentration of 25(OH)D in the blood plasma has established itself as the most suitable measure to assess whether a person is adequately supplied with vitamin D. An adequate concentration of 25(OH)D in the blood is absolutely necessary to ensure trouble-free synthesis of the biologically active metabolite 1,25(OH)$_2D_3$ both in the kidney (for endocrine effects) and at all its other synthesis sites (for autocrine/paracrine effects). The vitamin D formed in the skin under the influence of UV or ingested with food is rapidly and completely converted into 25(OH)D in the liver. The amount of 25(OH)D in the blood is thus the vitamin D storage or reservoir (*"How full is the tank?"*), which all organs access to meet their vitamin D needs.

The second metabolic step crucial for the conversion of 25(OH)D to the active vitamin D hormone 1,25(OH)$_2D_3$, hydroxylation, according to the classical view, mainly takes place in the kidney. The 1,25(OH)$_2D_3$ produced there, with the involvement of CYP27B1, is released into the blood, where, after binding to a transport protein (DBP), it can reach all other organs to exert so-called endocrine hormone effects. These mainly concern the regulation of bone and calcium metabolism. In advanced renal insufficiency, the synthesis of 1,25(OH)$_2D_3$ is significantly reduced. The production of 1,25(OH)$_2D_3$ in the kidney is strictly regulated by central control mechanisms, including the blood level of parathyroid hormone and other hormones, as well as the calcium and phosphorus serum level. If there is too much 1,25(OH)$_2D_3$ in the body, it is converted in the liver and other organs into biologically inactive metabolites (initially into 24,25(OH)$_2$D$_3$) by activating another cytochrome P450 enzyme (CYP24A1, vitamin D-24-hydroxylase). According to current knowledge, 1,25(OH)$_2$D$_3$ has a 100 to 1000 times higher biological activity than other known natural vitamin D metabolites. Measured by its biological effect, 1,25(CH)$_2$D$_3$ is probably the most active of the numerous naturally occurring substances that are formed from the "*mother substance*" vitamin D and are referred to as vitamin D metabolites (or also as calciferols). The exact designations (nomenclature) of the many different substances that belong to the vitamin D metabolism are precisely defined, but this is often not exactly applied. Thus, the term "vitamin D" is sometimes incorrectly used also for the biologically active metabolite 1,25(OH)$_2$D$_3$.

The autocrine/paracrine 1,25(OH)$_2$ D$_3$/VDR pathway
For understanding vitamin D metabolism, it is important to know that 1,25(OH)$_2$D$_3$ can reach its target cells via another metabolic pathway, referred to as autocrine/paracrine, in addition to the endocrine pathway. Over the past two decades, it has been shown that biologically active 1,25(OH)$_2$D$_3$ is not only produced in the kidney, but also in many other organs. It has been demonstrated that, for example, in the skin (including epithelial cells, keratinocytes) and in the immune system (including monocytes in the blood, T and B lymphocytes, and Langerhans cells in the skin), numerous cells not only have receptors for 1,25(OH)$_2$D$_3$ (vitamin D receptor, VDR) (which identifies these cells as target cells of vitamin D action), but also have the ability to locally synthesize 1,25(OH)$_2$ D$_3$ due to their enzyme equipment (CYP27B1). The 1,25(OH)$_2$D$_3$ produced in this way in many other organs, with the exception of the kidneys, is predominantly not released into the blood. Instead, it locally regulates different cell functions in these various organs by binding and activating the vitamin D receptor (VDR). This includes the regulation of cell growth, but also, for example, the control of insulin secretion in the pancreas. Unlike the endocrine system, this effect usually has no connection with bone and calcium metabolism. Rather, these autocrine/paracrine hormone effects control various functions, metabolic and growth processes in

numerous different cell types and organs, which are of great importance for the development and clinical course of numerous diseases, including infectious, metabolic, autoimmune and cancer diseases.

Endocrine and autocrine/paracrine pathways require different levels of 25(OH)D in the blood for optimal supply

Differences between the endocrine and the autocrine/paracrine pathway are also important for understanding the optimal vitamin D status in humans. In this context, it is important to know that both the endocrine and the autocrine/paracrine pathway rely on the same pro-hormone for the formation of the active metabolite $1,25(OH)_2D$, namely the biologically inactive storage form of vitamin D present in the blood and bound to transport proteins, 25(OH)D. Research results from recent years have shown that both pathways require different blood concentrations of 25(OH)D to ensure their optimal effect. Only the kidney has special proteins (i.e., megalin, cubilin), which facilitate the uptake of the prohormone 25(OH)D. Therefore, relatively low 25(OH)D serum levels (>10 ng/ml) are sufficient for it to produce adequate amounts of the biologically active metabolite $1,25(OH)_2D$ for the endocrine pathway. In contrast, other organs require significantly higher 25(OH)D concentrations in the blood (probably >20 ng/ml) to produce the active metabolite $1,25(OH)_2D$ in sufficient quantity to ensure an optimal autocrine/paracrine $1,25(OH)_2D$ effect in these tissues.

The classic vitamin D receptor (VDR): a nuclear transcription factor that regulates more than 1000 coding genes

Biologically active vitamin D metabolites mediate a significant part of their effect through the activating binding to corresponding receptors in target cells. The important function of the classic vitamin D receptor (VDR) for mediating the endocrine and autocrine/paracrine effects of $1,25(OH)_2D$ has been known for many years. The $1,25(OH)_2 D_3$/VDR complex binds in the cell nucleus to specific DNA sequences in the promoter region of its target genes. This regulates their reading (transcription) and the subsequent protein synthesis. Simply put, these genes are switched on and off, which subsequently leads to different biological effects. Vitamin D regulates more than 1000 of the approximately 20,000 coding genes of the human organism, thus about 5% of the genome, our entire genetic information.

New vitamin D pathways that are not mediated by $1,25(OH)_2$ D and VDR, but by other biologically active metabolites and their corresponding receptors

In recent years, numerous new vitamin D pathways have been identified, the effects of which are not mediated by the classical VDR, but by other receptor molecules. It has recently been shown that various vitamin D metabolites such as 20,23-dihydroxyvitamin D ($20,23(OH)_2D_3$), which were previously attributed only a subordinate importance, also exert important biological effects. Thus, 20,23-dihydroxyvitamin D and other vitamin D metabolites bind to aryl hydrocarbon receptor (AhR), peroxisome proliferator-activated receptors (PPARs), liver X receptors (LXRs) and so-called retinoic acid receptor-related orphan receptors (RORs). This regulates numerous other important signaling pathways in addition to our *'internal clock'*, which are also involved in the development of many cancer and metabolic diseases. As a so-called "dioxin receptor", the AhR is of central importance for the detection and detoxification of environmental toxins.

Vitamin D pathways—a strictly regulated network for controlling/fine-tuning various metabolic functions

There are many reasons why vitamin D can exert so many different effects. These include, in addition to several different mechanisms of action, the multitude of different active metabolites (including $1,25(OH)_2$ D_3, $20,23(OH)_2D_3$), their various formation pathways and different distribution (e.g., autocrine/paracrine) as well as their numerous corresponding specific receptor proteins (including VDR, LXR, AhR, PPARγ, RORs) and their different distribution in target cells. This ensures that the required hormone effect can be individually dosed and locally adapted to the respective desired target genes and physical needs through various regulatory mechanisms.

Oral intake (supplementation) of vitamin D does not replace all positive effects of UV-induced vitamin D synthesis in the skin
It is interesting to note that biologically active vitamin D metabolites have been detected not only above the vitamin D synthesis (e.g., 20,23(OH)$_2$D), but also before it (including tachysterol derivatives). This suggests that oral intake (supplementation) of vitamin D cannot compensate for all the positive effects of UV-induced vitamin D synthesis in the skin.

1.3 Misconception 3: *There is Scientific Evidence Refuting the Great Importance of Vitamin D for Health*

1.3.1 Correction

There is no scientific evidence refuting the relevance of Vitamin D for health.

1.3.2 Comment

1.3.2.1 What to Consider When Interpreting Scientific Studies

Publications that claim or suggest such a connection have major weaknesses, even if they were published in renowned scientific journals. As an example, the publication by Manson and colleagues (Manson et al. 2019) is mentioned, which wanted to investigate the benefit of Vitamin D in the primary prevention of cancer and cardiovascular diseases. For this purpose, over 25,000 subjects were included in a randomized intervention study (VITAL study) and examined over a period of more than 5 years. In the group of subjects who had taken 2000 units of Vitamin D in addition to Omega-3 fatty acid, no lower rate of new cancer and fatal cardiovascular diseases was seen. At first glance, this study should be quite suitable to gain important insights to clarify the question. Upon closer inspection, however, it is completely unsuitable for this. The decisive weakness of the work is that the average of the 25(OH)D serum value in the cohort studied was 30.8 ng/ml. Thus, the Vitamin D status was far above the critical Vitamin D deficiency range (25(OH)D serum value <20 ng/ml). Thus, this study only confirms the safety of oral substitution with 2000 IU Vitamin D in individuals with sufficient 25(OH)D serum level. However, statements about the effects of Vitamin D deficiency cannot certainly be derived from this. Because such a collective is completely unsuitable to capture the effects of compensating a Vitamin D deficiency (approx. 60% of the German population even have a Vitamin D deficiency/25(OH)D serum values < 20 ng/ml in summer).

Another example is the meta-analysis by Keum et al. (2019). This comprehensive analysis of the literature includes both studies whose participants had too high 25(OH)D serum values at the beginning of the study, as well as studies that administered too low doses of Vitamin D (400 IU) daily.

▸ **Note**
There is no scientific evidence refuting the great importance of Vitamin D for our health. Publications that question such a connection have major weaknesses, even if they were published in renowned scientific journals.

References

Ang Z, Selvanayagam JB, Hyppönen E. Non-linear mendelian randomization analyses support a role for vitamin D deficiency in cardio vascular disease risk. Eur Heart J. 2021. https://doi.org/10.1093/eurheartj/ehab809.

Keum N et al. Vitamin D supplementation and total cancer incidence and mortality: a meta-analysis of randomized controlled trials. Ann Oncol. 2019;30(5):733–43.

Kuznia S, Zhu A, Akutsu T, Buring JE, Camargo CA Jr, Cook NR, Chen LJ, Cheng TD, Hanturen S, Lee IM, Manson JE, Neale RE, Scragg R, Shadyab AH, Sha S, Sluyter J, Tuomainen TP, Urashima M, Virtanen JK, Voutilainen A, Wactawski-Wende J, Waterhouse M, Brenner H, Schöttker B. Efficacy of vitamin D_3 supplementation on cancer mortality: systematic review and individual patient data meta-analysis of randomised controlled trials. Ageing Res Rev. 2023. https://doi.org/10.1016/j.arr.2023.101923.

Manson JE et al. Vitamin D supplements and prevention of cancer and cardiovascular disease. N Engl J Med. 2019;380(1):33–44.

Martineau AR, Jolliffe DA, Hooper RL, Greenberg L, Aloia JF, Bergman P, Dubnov-Raz G, Esposito S, Ganmaa D, Ginde AA, Goodall EC, Grant CC, Griffiths CJ, Janssens W, Laaksi I, Manaseki-Holland S, Mauger D, Murdoch DR, Neale R, Rees JR, Simpson S Jr, Stelmach I, Kumar GT, Urashima M, Camargo CA Jr. Vitamin D supplementation to prevent acute respiratory tract infections: systematic review and meta-analysis of individual participant data. BMJ. 2017;15(356):i6583. https://doi.org/10.1136/bmj.i6583. PMID: 28202713; PMCID: PMC5310969.

Munger KL, Levin LI, Hollis BW, Howard NS, Ascherio A. Serum 25-hydroxyvitamin D levels and risk of multiple sclerosis. JAMA. 2006;296(23):2832–8. https://doi.org/10.1001/jama.296.23.2832. PMID: 17179460.

Niedermaier T, Gredner T, Kuznia S, Schöttker B, Mons U, Lakerveld J, Ahrens W, Brenner H. Vitamin D food fortification in European countries: The underused potential to prevent cancer deaths. Eur J Epidemiol. 2022. https://doi.org/10.1007/s10654-022-00867.

Plewig G et al. Braun-Falco's Dermatologie. 6. Aufl. Berlin/Heidelberg/New York: Venerologie und Allergologie/Springer; 2016.

Reichrath J, Herausgeber. Sonne – die Dosis macht's! Berlin/Heidelberg: Springer; 2021.

Wacker M, Holick MF. Sunlight and Vitamin D: A global perspective for health. Dermatoendocrinol. 2013;5(1):51–108. https://doi.org/10.4161/derm.24494. PMID: 24494042; PMCID: PM C3897598.

Zittermann A, Hrsg. Vitamin D im Überblick. Berlin/Heidelberg: Springer; 2022a.

Zittermann A. Vitamin D im Überblick. Essentials. Berlin/Heidelberg: Springer Spektrum; 2022b. https://doi.org/10.1007/978-3-662-65716-4 17. Abb 2.1.

Vitamin D Intake with Food

Contents

2.1 Misconception 1: *Enriching Food With Vitamin D Has No Impact on Health and Therefore Makes No Sense* 33
2.2 Mistake 2: *Under Our Living Conditions, the Vitamin D Requirement Is Fully Covered by the Diet—Even Without Taking Dietary Supplements* 37
2.3 Misconception 3: *It is a Flaw in Evolution that Breast Milk Contains Everything Infants Need for Healthy Development—With the Exception of Vitamin D* 40
2.4 Misconception 4: *It is Necessary to take Vitamin D Together with Fatty Foods to Achieve Good Absorption in the Gastrointestinal Tract* 42
References 43

2.1 Misconception 1: *Enriching Food With Vitamin D Has No Impact on Health and Therefore Makes No Sense*

2.1.1 Correction

It is assumed that the enrichment of food with vitamin D would have a very positive impact on the health of the German or European population.

© The Author(s), under exclusive license to Springer-Verlag GmbH, DE, part of Springer Nature 2025
J. Reichrath, *Vitamin D in Focus*,
https://doi.org/10.1007/978-3-662-71341-9_2

2.1.2 Comment

2.1.2.1 Only a Few Foods Contain Significant Amounts of Vitamin D

Most of our food contains only small amounts of vitamin D (Table 2.1). The enrichment of food could significantly improve the vitamin D supply of our population. In many industrialized countries, food was successfully enriched with vitamin D in the 1930s and 1940s. As a result, for example in the USA, the prevalence of rickets decreased from 50% to 0.5%. In Germany, only a few foods, including margarine and spreads, are currently legally allowed to be supplemented with vitamin D. However, the addition of vitamin D is also permissible by general order or exemption, such as in cooking oil, breakfast cereals, plant cream or fresh cheese preparations. If the requirements of the Novel Food Regulation are met, ergosterol- or cholesterol-containing foods such as yeast, bread or milk can also be enriched with vitamin D by UV irradiation. Industrially produced infant milk powder, which is legally considered a dietary product, is also enriched with vitamin D (often 10 µg per liter of final product), which is referred to as silent rickets prophylaxis. A successful example of a broad-based vitamin D enrichment of food is Finland, where voluntary food enrichment was introduced in 2003. It was recommended to add vitamin D in a dose of 10 µg/100 g to all spreads and in a dose of 0.5 µg/100 g to all liquid milk products. In 2010, these enrichment recommendations were doubled. A representative nationwide survey of the population showed that the average 25(OH)D concentrations in serum increased from 47.6 nmol/L in 2000 to 65.4 nmol/L in 2011. The prevalence of 25(OH)D concentrations below 30 or 50 nmol/L thus decreased from 13.0 or 55.7% (2000) to 0.6 or 9.1% (2011).

2.1.2.2 The Systematic Enrichment of Food with Vitamin D Could Prevent Many Deaths

Scientists from the German Cancer Research Center (DKFZ) recently determined using statistical model calculations that the systematic enrichment of food with vitamin D in Europe

Table 2.1 Vitamin D content of selected foods/organisms

Food/Organism	Micrograms (μg) per 100 g	International Units (IU) per 100 g
Sea creatures		
Eel, smoked	22	880
Herring, fresh	6–25	240–1000
Wild salmon, fresh	6–25	240–1000
Farmed salmon, fresh	6	240
Tuna	3–7	120–280
Mackerel	8–16	320–640
Sardine, in oil	4	160
Cod liver oil, traditional	210–330	8400–13.200
Seal, adipose tissue	75	3000
Beluga whale, adipose tissue	43	1720
Polar bear, adipose tissue	40	1600
Dairy products, eggs		
Whole milk	0.2	8
Infant milk, industrially produced, Germany	t	
Butter	1	40
Gouda	1	40
Parmesan	0.7	28
Eggs	3	120
Meat		
Beef liver	1	40
Chicken	0.1–2.0	4–80
Turkey	2	80
Pork	0.7–1.5	28–60
Beef	0.1–1.0	4–40
Mushrooms		
Champignons, fresh	2	80
Morels, fresh	3	120
Shiitake, fresh	2	8

could prevent more than 100,000 cancer-related deaths per year (Kuznia et al. 2023). Vitamin D deficiency is not only associated with bone and muscle diseases, but also with numerous other diseases (Reichrath 2021; Wacker and Holick 2013). Meta-analyses of large randomized studies have shown that taking vitamin D supplements reduces cancer mortality rates by about 13%. Although most foods contain only a small amount of vitamin D (Table 2.1), consuming foods enriched with vitamin D increases vitamin D levels in a similar way to taking vitamin D supplements. Some countries, including the USA, Canada, and Finland, have been enriching food with vitamin D for some time (Pilz et al. 2018). However, most other nations have not yet done so.

Epidemiologists at the German Cancer Research Center (DKFZ) investigated the possible influence of targeted enrichment of food with vitamin D on cancer mortality in Europe. They first collected information on the guidelines for food supplementation with vitamin D from 34 European countries. In addition, the scientists determined the number of cancer-related deaths and life expectancy in each country from databases. They linked this information with the results of studies on the influence of vitamin D administration on cancer mortality rates. Using statistical methods, they estimated the number of cancer-related deaths that are already being prevented in countries with food enrichment. They also calculated the number of deaths that could be additionally avoided if all European countries were to introduce the enrichment of vitamin D in food.

The researchers concluded that vitamin D enrichment currently prevents about 27,000 cancer deaths in all considered European countries per year. They estimated in a model calculation that about 130,000 or about 9% of all cancer deaths in Europe could be prevented if all countries considered in this study were to enrich food with adequate amounts of vitamin D. According to the authors, such a measure could save almost 1.2 million years of life. The researchers also pointed out that adequate vitamin D supply can be ensured through sun exposure in addition to dietary intake. The DKFZ's Cancer Information Service recommends spending about 12 minutes outdoors two to three times a week when the sun is shining. The face, hands, and parts of the arms and legs should be exposed to the sun's rays uncovered and without protection for this period of time.

▶ **Note** It is expected that the enrichment of food with vitamin D would have a very positive impact on the health of the German and European population.

2.2 Mistake 2: *Under Our Living Conditions, the Vitamin D Requirement Is Fully Covered by the Diet—Even Without Taking Dietary Supplements*

2.2.1 Correction

Under our living conditions, it is very difficult and only theoretically possible to meet the need for vitamin D completely by adjusting the diet (Reichrath 2021; Wacker and Holick 2013).

2.2.2 Comment

2.2.2.1 Diet: of Minor Importance for Vitamin D Supply

We can meet our need for vitamin D in two completely different ways. Vitamin D is produced both in the skin under the influence of UV-B rays from precursors and is also taken in with food. According to studies, under our living conditions, about 80–95% of the required vitamin D must be formed in the skin with the involvement of UV-B radiation. This is because the skin is the only organ in humans capable of producing vitamin D (Reichrath 2021; Wacker and Holick 2013). Only a few foods (including fatty fish such as salmon, mackerel and herring, fish oils, cod liver oil, offal, eggs, mushrooms and to a limited extent also dairy products) contain relevant amounts of vitamin D. The low vitamin D content of most foods (Table 2.1) practically does not allow us to achieve sufficient vitamin D supply by adjusting our diet under our living conditions. As a rule, under our dietary conditions ("western diet"), only about 5–20% of the requirement can be covered by intake with food (Saternus et al. 2019).

The German Society for Nutrition (DGE) regularly publishes estimates on the actual and desired daily intake of vitamin D through food. According to their 14th nutrition report, in the absence of endogenous vitamin D synthesis in the skin, the estimated value for the intake of vitamin D required by every person in Germany is about 20 µg (800 IU) daily. However, the actual daily intake of vitamin D through food is much lower. In Germany, according to DGE estimates, it is only 1.1–2.2 µg. Therefore, an "enrichment" of certain foods with vitamin D, as has been successfully practiced in the USA with milk, orange juice, and many other foods for many decades, is urgently needed for Germany/Europe. Dietary supplements are also well suited and important to prevent a vitamin D deficiency. The advantage of taking supplements is that they can be precisely dosed and are cost-effective. In addition, they are freely available in the required dosage in both drugstores and pharmacies, safe, and always available. Depending on the dosage and purpose of use, vitamin D preparations are classified as a supplement or medicine on a case-by-case basis. According to the opinion of a joint expert commission of the Federal Office for Consumer Protection and Food Safety and the Federal Institute for Drugs and Medical Devices, vitamin D preparations can be classified as dietary supplements up to a dose of 20 µg (800 IU)/day, as they aim to maintain an adequate vitamin D supply. However, the labeling must not justify classification as a medicine. According to the statement of this expert commission, a dosage above 20 µg (800 IU)/day does not necessarily justify classification as a medicine. Both daily and weekly intake of vitamin D should ensure adequate vitamin D supply. However, many experts consider daily intake of vitamin D to be more favorable than higher-dose weekly intake. One reason is that with the weekly intake of vitamin D in higher doses, degrading proteins (enzymes) are activated more than with the lower-dose daily intake. Thus, the dose-effect relationship with the lower-dose daily intake of vitamin D is probably more favorable than with the weekly intake as a higher-dose bolus.

2.2 Mistake 2: *Under Our Living Conditions* …

For decades in Germany, the so-called continuous rickets prophylaxis (as opposed to the previous high-dose bolus administration) has been successfully carried out in infants with daily oral doses of 10–12.5 µg (400–500 IU). Since the preparations explicitly serve to prevent disease, they are legally classified as medicines and are prescribed by a doctor. Special care should be taken to ensure the vitamin D supply of pregnant women, according to recommendations from the German network "Healthy into Life—Young Family Network", as they have an increased risk of vitamin D deficiency. The intake of a supplement is also advisable for other risk groups for a vitamin D deficiency (e.g., low sun exposure of the skin due to short stays outdoors, covering clothing or use of sun creams, dark skin, and fat malabsorption). In Switzerland, the Federal Office of Public Health recommends not only for pregnant and breastfeeding women, but also for children under 3 years of age, a vitamin D intake of 15 µg/day in addition to the vitamin D administration to infants of 10 µg/day. The recommendations of the British National Institute for Health and Care Excellence (NICE) also emphasize the importance of a wide availability of vitamin D supplements not only for pregnant and breastfeeding women, but also for children up to the age of 5. In the absence of skin synthesis, the D-A-CH (Germany-Austria-Switzerland) nutrition societies recommend the intake of a vitamin D supplement of 20 µg daily for all age groups > 1 year. In nursing homes, the administration of vitamin D can now be considered mandatory. Nowadays, vegans can also get plant-based vitamin D. This is obtained from mushrooms or lichens by converting ergosterol into vitamin D_2 through UV-B irradiation.

▶ **Note** We can meet the need for vitamin D in two different ways: through UV-induced production in the skin and through intake with food. Since only a few foods contain significant amounts of vitamin D, under our dietary and living conditions, we need to produce 80–90% of the required vitamin D in the skin.

2.3 Misconception 3: *It is a Flaw in Evolution that Breast Milk Contains Everything Infants Need for Healthy Development—With the Exception of Vitamin D*

2.3.1 Correction

New scientific findings suggest that the low vitamin D content in breast milk is not a flaw in evolution. The cause is rather the vitamin D deficiency of breastfeeding mothers.

2.3.2 Comment

2.3.2.1 Infants, Vitamin D and Sun: What Do Professional Societies Recommend?

Many professional societies recommend that infants (both light and dark-skinned) should generally not be exposed to direct sunlight, as their skin's own UV protection mechanism is not yet mature. Due to the low exposure to sunlight, infants can only produce a small amount of vitamin D in their skin. Since under our living conditions breast milk contains only a small amount of vitamin D and the vitamin D content of industrially produced infant food (milk) is also very low (breast milk contains on average 0.073 µg vitamin D per 100 ml [Souci et al. 2008], in pre- and 1-food the vitamin D content is about 0.9–1.2 µg per 100 ml of ready-to-drink food [D-A-CH 2012]), infants are considered a risk group for vitamin D deficiency according to usual assessments. According to the recommendations of many professional societies, including the "Healthy into Life" network, an IN FORM initiative of the Federal Ministry of Nutrition and the participating professional societies (German Society for Nutrition [DGE], German Society for Child and Adolescent Medicine [DGKJ], Research Institute for Child

2.3.2.2 Does Breast Milk Contain Enough Vitamin D?

Nutrition [FKE]), every infant should therefore receive 400–500 International Units (IU; equivalent to 10–12.5 µg) of vitamin D daily through drops or tablets.

The American gynecologist Dr. Bruce Hollis (Hollis and Wagner 2017) asked himself the following interesting question a few years ago: How can it be that Mother Nature has created a "super elixir" with breast milk that provides the infant with everything he needs for life, but there is one hardly understandable exception: It contains too little vitamin D. Because breast milk usually contains the very small amount of less than 50 IU of vitamin D per liter. This is very surprising. Because a sufficient supply of vitamin D is particularly important for infants, also to ensure the integrity of the developing bone skeleton and to prevent the bone disease rickets. Dr. Hollis then gave mothers of fully breastfed infants vitamin D in a relatively high dosage (6400 IU daily). He reports in a study published in 2015 that under these conditions breast milk contains enough vitamin D to adequately supply the infants with vitamin D. The results show that a sufficient transfer of vitamin D from the mother to the breast milk occurs from 25(OH)D serum levels of the mother of about 48 ng/ml. Dr. Hollis convincingly explains the low vitamin D content of breast milk (which in the past also led to rickets in breastfed children) by the widespread vitamin D deficiency of breastfeeding mothers. He concludes that Mother Nature has indeed intended that breast milk contains enough vitamin D. The reason our infants need a supplemental supply of vitamin D is because their mothers have a vitamin D deficiency.

▶ **Note** The reason our infants need a supplemental supply of vitamin D is because their mothers have a vitamin D deficiency.

2.4 Misconception 4: *It is Necessary to take Vitamin D Together with Fatty Foods to Achieve Good Absorption in the Gastrointestinal Tract*

2.4.1 Clarification

It has been reported that it is not important for absorption in the intestine whether Vitamin D is taken together with fatty foods. A study showed no difference between simultaneous intake of Vitamin D with low-fat fruit juice or with a fatty biscuit.

2.4.2 Comment

2.4.2.1 How is Vitamin D Absorbed in the Intestine?

Vitamin D absorbed (exogenous) in the fat phase of food is absorbed 60 to 100% in the small intestine (duodenum and jejunum). The involvement of bile acids, especially deoxycholic acid, is important. The transport into the lymphatic system (thoracic duct) and then from there into the vascular system is initially carried out in chylomicrons and particles from lipoproteins (especially low-density lipoproteins [LDL]). A study showed that for the efficiency of absorption of Vitamin D_2 or D_3 in the small intestine it hardly makes any difference whether both substances are taken with acidic fruit juice or with a fatty biscuit. This observation contradicts the frequently expressed opinion that Vitamin D supplements should be taken together with fatty foods to achieve effective absorption in the gastrointestinal tract. Patients with inflammatory diseases of the small intestine, with liver or bile dysfunction often show a reduced absorption rate of Vitamin D and therefore have an increased risk of developing a Vitamin D deficiency (Reichrath 2021; Wacker and Holick 2013).

▶ **Note** It hardly makes any difference for absorption in the intestine whether Vitamin D is taken together with fatty foods.

References

Hollis BW, Wagner CL. New insights into the vitamin D requirements during pregnancy. Bone Res. 2017;29(5):17030. https://doi.org/10.1038/boneres.2017.30. PMID: 28868163; PMCID: PMC5573964.

Kuznia S, Zhu A, Akutsu T, Buring JE, Camargo CA Jr, Cook NR, Chen LJ, Cheng TD, Hantunen S, Lee IM, Manson JE, Neale RE, Scragg R, Shadyab AH, Sha S, Sluyter J, Tuomainen TP, Urashima M, Virtanen JK, Voutilainen A, Wactawski-Wende J, Waterhouse M, Brenner H, Schöttker B. Efficacy of vitamin D3 supplementation on cancer mortality: Systematic review and individual patient data meta-analysis of randomised controlled trials. Ageing Res Rev. 2023. https://doi.org/10.1016/j.arr.2023.101923.

Pilz S, März W, Cashman KD, Kiely ME, Whiting SJ, Holick MF, Grant WB, Pludowski P, Hiligsmann M, Trummer C, Schwetz V, Lerchbaum E, Pandis M, Tomaschitz A, Grübler MR, Gaksch M, Verheyen N, Hollis BW, Rejnmark L, Karras SN, Hahn A, Bischoff-Ferrari HA, Reichrath J, Jorde R, Elmadfa I, Vieth R, Scragg R, Calvo MS, van Schoor NM, Bouillon R, Lips P, Itkonen ST, Martineau AR Lamberg-Allardt C, Zittermann A. Rationale and plan for vitamin D food fortification: a review and guidance paper. Front Endocrinol (Lausanne). 2018;17(9):373. https://doi.org/10.3389/fendo.2018.00373. PMID: 30065699; PMCID: PMC6056629.

Reichrath J, Herausgeber. Sonne – die Dosis macht's! Berlin/Heidelberg: Springer; 2021.

Saternus R, Vogt T, Reichrath J. A critical appraisal of strategies to optimize vitamin D status in Germany, a population with a western diet. Nutrients. 2019;11(11):2682. https://doi.org/10.3390/nu11112682. PMID: 31698703; PMCID: PMC6893762.

Souci SW, Fachmann W, Kraut H. Die Zusammensetzung der Lebensmittel. Nährwert-Tabellen, 7., revid. u. erg. Aufl., Wissenschaftliche Verlagsgesellschaft mbH; 2008, Stuttgart.

Wacker M, Holick MF. Sunlight and Vitamin D: a global perspective for health. Dermatoendocrinol. 2013;5(1):51–108. https://doi.org/10.4161/derm.24494. PMID: 24494042; PMCID: PMC3897598.

UV-Induced Vitamin D Production in the Skin

3

Contents

3.1	Misconception 1: *Vitamin D Can Be Poduced in the Skin Without UV Influence*	46
3.2	Misconception 2: *Natural Sun Exposure Allows for Sufficient Vitamin D Production in Germany in All Seasons*	47
3.3	Misconception 3: *The Vitamin D Production of the Skin is Only Insignificantly Influenced by the Strength of the Ground-Level UV Radiation*	49
3.4	Misconception 4: *Skin's Vitamin-D Synthesis is Not Influenced by Age*	53
3.5	Misconception 5: *The Synthesis of Vitamin D Lowers the Blood Cholesterol Level by Consuming Its Precursor 7-DHC*	53
3.6	Misconception 6: *The Skin Produces Only a Few Hundred IU of Vitamin D Daily, Even Under Sufficient UV-B Influence*	54
3.7	Misconception 7: *A Healthy Person Can Produce Too Much Vitamin D From Extensive Sunbathing and Thereby Endanger Their Health*	56
3.8	Misconception 8: *UV-B Irradiation of Dairy Cows Has No Effect on the Vitamin D Content of Cow's Milk*	58
3.9	Misconception 9: *UV Irradiation of Food Has No Effect on Its Vitamin D Content*	59
3.10	Misconception 10: *The Impact of Changes in Our Vitamin D Supply on Vitamin D Status is Difficult to Estimate*	60
References		62

© The Author(s), under exclusive license to Springer-Verlag GmbH, DE, part of Springer Nature 2025
J. Reichrath, *Vitamin D in Focus*,
https://doi.org/10.1007/978-3-662-71341-9_3

3.1 Misconception 1: *Vitamin D Can Be Produced in the Skin Without UV Influence*

3.1.1 Correction

UV-B radiation is essential for the formation of vitamin D in the skin (Reichrath 2021; Wacker and Holick 2013).

3.1.2 Comment

3.1.2.1 The Skin's Vitamin D Biosynthesis

The starting point of the vitamin D biosynthesis (Fig. 1.1) in the skin is 7-dehydrocholesterol (7-DHC), a so-called sterol and precursor of cholesterol (Fig. 1.1). 7-DHC is produced directly in the skin in a multi-step process and stored there in large quantities (about 2000 ng/cm^2) in the outer cell membranes. Through a photochemical reaction known as photolysis, the B-ring is initially broken open under the influence of UV-B (280–315 nm, Fig. 3.1) radiation and 7-DHC in the skin cell membranes is converted into the thermodynamically unstable previtamin D_3 (Reichrath 2021; Wacker and Holick 2013). The action spectrum of UV rays in the conversion of 7-DHC to previtamin D peaks at about 298 nm. Through a subsequent time- and temperature-dependent chemical reaction (a *"sigmatropic shift"* of a proton from C-19 to C-9 with subsequent isomerization), previtamin D is converted into V*itamin D_3*. After that, the vitamin D produced in the skin (endogenously) under UV influence is released into the blood within 8 to 72 hours.

▶ **Note** The energy of UV-B radiation is essential for the skin's synthesis of vitamin D.

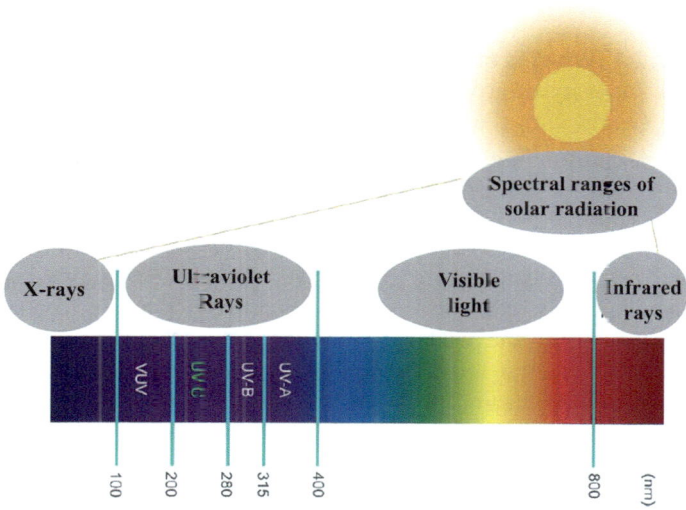

Fig. 3.1 Spectral ranges of solar radiation. In addition to ultraviolet (UV) rays, infrared rays, and visible light, the sun can also emit X-rays from its outermost atmospheric layer (corona), which averages 2 million degrees Celsius. Exposure of the skin to UV-B (280–315 nm) rays is crucial for cutaneous vitamin D formation

3.2 Misconception 2: *Natural Sun Exposure Allows for Sufficient Vitamin D Production in Germany in All Seasons*

3.2.1 Correction

Due to the shallow angle of incidence, the sun's rays in Central Europe have a long path to the earth during the winter months (October–March). Since the atmosphere filters UV-B radiation, not enough of these rays reach the earth's surface to ensure sufficient vitamin D production in winter (the "*vitamin D winter*").

3.2.2 Comment

3.2.2.1 What is meant by the vitamin D winter?

The intensity of ground-level UV-B radiation in Germany varies greatly with the seasons. Due to its geographical location between the 47th and 54th northern latitude, this radiation peaks in summer and reaches a minimum in winter. The atmosphere only allows a portion of the solar UV radiation to pass through, with this filtering effect increasing with the length of the path and particularly affecting short-wave UV rays. Therefore, not enough UV-B radiation reaches the earth's surface in Central Europe during the winter months (October–March) to ensure sufficient vitamin D production (the *"vitamin D winter"*). For vitamin D production in the skin, sufficient sun exposure with UV-B radiation of a certain minimum intensity is absolutely necessary (Reichrath 2021; Wacker and Holick 2013). This minimum intensity of solar radiation for vitamin D synthesis is thus only exceeded in Germany for about 6 months of the year (when a UV index of 3 is reached). Therefore, sufficient vitamin D synthesis is only possible during this period, with other factors also being important. For example, the UV-B intensity is strongest between 12 and 3 p.m. (when the sun is at its zenith and its rays have the shortest path through the atmosphere to the earth; Figure 3.2) and is low even in summer before 9 a.m. and after 5 p.m.

▶ **Note** Natural sun exposure does not allow for sufficient vitamin D production in Germany in all seasons. This is because, due to the shallow angle of incidence of the sun's rays, not enough UV-B radiation reaches the earth's surface in Central Europe during the winter months (October–March) after the long path through the atmosphere to ensure sufficient vitamin D production (the "vitamin D winter").

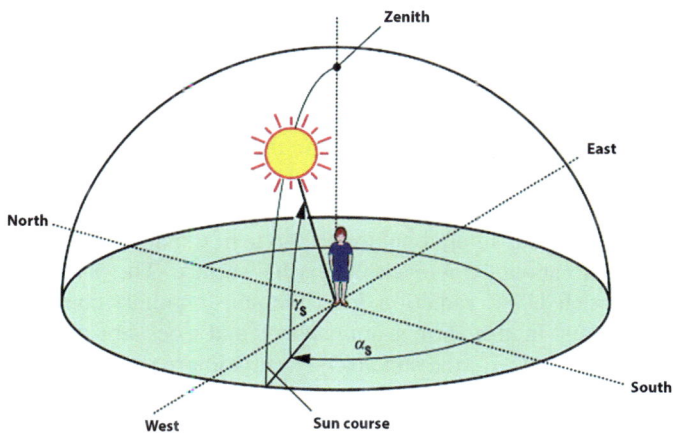

Fig. 3.2 Sun position and cutaneous vitamin D production. The term sun position refers to the angle between the sun's rays and an imaginary vertical axis at the earth's surface. For example, this angle is 0 degrees when the sun is at its highest point in the zenith. The lower the sun is, for example in the late afternoon or in winter, the larger this angle becomes. With the associated increase in the distance to be covered through the atmosphere, the proportion of the sun's UV radiation that reaches the earth's surface decreases. The short-wave UV-B radiation is more affected by this decrease than the longer-wave UV-A radiation. (*Source*: Wacker and Holick 2013, with kind permission from Dr. Holick)

3.3 Misconception 3: *The Vitamin D Production of the Skin is Only Insignificantly Influenced by the Strength of the Ground-Level UV Radiation*

3.3.1 Correction

Factors that influence the intensity of ground-level UV-B radiation also change the UV-induced production of vitamin D in the skin.

3.3.2 Comment

3.3.2.1 *What Influences Ground-Level UV Radiation?*

How much vitamin D is produced in the skin? The answer to this question largely depends on the available solar UV-B radiation (Fig. 3.1) (Reichrath 2021; Wacker and Holick 2013). Since a large part of it is filtered out in the ozone of the high atmosphere, only up to about 10% reach the earth's surface. The particularly energy-rich UV-C radiation from the sun is almost completely filtered out in the earth's atmosphere and does not reach the ground. In contrast, the sun's UV-A radiation reaches the earth's surface almost unhindered, as it largely passes through the atmosphere unfiltered due to its relatively long wavelength.

The basic rule always applies: The longer the path of solar radiation through the atmosphere, the lower the proportion of ground-level UV-B radiation. When the sun is high (Fig. 3.2), the path through the atmosphere to the Earth's surface is relatively short, which is why a large proportion of UV-B radiation reaches the Earth's surface. Therefore, the ground-level UV-B radiation is strongest when the sun is high. The time of day determines the angle of incidence of sunlight and thus also regulates the length of the path through the atmosphere to the Earth's surface. During the course of the day, the radiation is stronger at noon than in the morning or evening. From 11 to 15 o'clock, a particularly high proportion of UV-B radiation reaches the Earth's surface. In our relatively moderate middle latitudes, about 75% of the total daily ground-level UV-B radiation reaches the ground in summer between 9 and 15 o'clock, and the rule of thumb is that about 50% of the erythemally effective energy radiates onto the Earth's surface from one hour before to one hour after solar noon. The season also determines the angle of incidence of sunlight and thus the length of the path through the atmosphere to the Earth's surface. The radiation is stronger in summer than in winter. At the equator, the energy of the average UV radiation is more than a thousand times stronger than at the North and South Poles. With increasing altitude of the location above sea level, the energy of the UV radiation increases, by about 10–20% per km of altitude increase. Other location-specific factors such as reflection of radiation from the ground as well as scattering and absorption by water also influence the strength of natural UV radiation at the Earth's surface.

The scattering of rays is of great importance for the effect of the atmosphere as a UV filter. Since scattering also strongly depends on the wavelength, the shorter-wave UV-B radiation is more strongly attenuated in this process than the longer-wave UV-A radiation. Thus, clouds usually weaken UV-B radiation by scattering on the water droplets. With strong cloud cover, the UV-B radiation is therefore much weaker than with clear skies. UV-A radiation is hardly stopped by thin cloud cover, certain cloud situations can even increase the UV exposure through scattered radiation. However, the strength of UV radiation is often underestimated even with light cloud cover. Water, sand, and snow also scatter and reflect the radiation. Even 0.5 m below the water surface, the UV radiation still reaches 40% of the intensity at the water surface. Snow reflects up to 80%, bright sand up to 25% of the UV radiation. This can even increase the radiation exposure. Shade usually significantly reduces the UV exposure. Like the ozone in the stratosphere, ground-level ozone from the exhaust gases of industry and car traffic generated by summer smog also influences the intensity of ultraviolet radiation at the Earth's surface. Only an average of 50% of the total ground-level UV radiation comes directly from the sun. The other half is emitted by indirect radiation from the blue sky. This means that the total amount of ground-level UV radiation can also be significantly determined by the surrounding situation even with direct sunlight (Fig. 3.3). Shielding of the indirect rays due to surrounding mountains, trees or the degree of development can cause a significant weakening of natural UV radiation. On the other hand, this also means that shade often provides insufficient protection against sunburn when there is a wide horizon at the same time. A typical example of such insufficient protection is staying under a small sun umbrella on the beach.

It should be noted that in addition to these general factors such as geographical latitude, altitude, season, time of day, weather, air pollution and surface reflection, individual factors

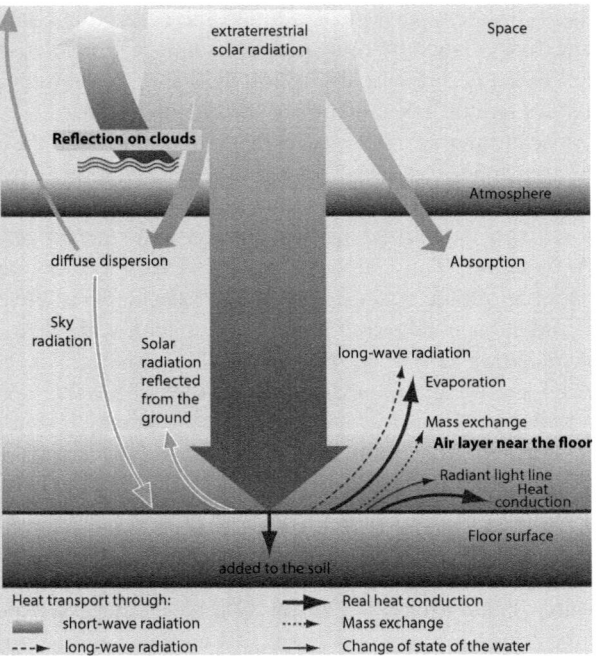

Fig. 3.3 Factors influencing the sun's rays on their way to Earth. (*Source*: Krause and Stange 2012)

such as skin type/color, age, clothing, use of skin/sun creams, body weight, the constant decrease in outdoor stay time as well as other influences of modern lifestyle affect the skin's exposure to UV-B and thus also the skin's vitamin D production (Fig. 3.3).

▶ **Note** In addition to general factors such as geographical latitude, altitude, season, time of day, weather, air pollution and surface reflection, individual factors such as skin type/color, age, clothing, use of skin/sun creams, body weight and outdoor stay time also influence the skin's exposure to UV-B and thus its vitamin D production.

3.4 Misconception 4: *Skin's Vitamin-D Synthesis is Not Influenced by Age*

3.4.1 Correction

The truth is that UV-induced Vitamin-D synthesis in the skin decreases with age

3.4.2 Comment

3.4.2.1 What Influence Does Age Have on the Skin's Vitamin-D Production?

The 7-Dehydrocholesterol content (7-DHC) of the skin decreases with age. It is estimated that—partly due to the reduced 7-DHC content—in humans in old age (approx. >60 years) the skin's ability to form Vitamin D_3 decreases by about a factor of 3 (Reichrath 2021; Wacker and Holick 2013).

▶ **Note** UV-induced Vitamin-D synthesis in the skin decreases with age.

3.5 Misconception 5: *The Synthesis of Vitamin D Lowers the Blood Cholesterol Level by Consuming Its Precursor 7-DHC*

3.5.1 Correction

The formation of Vitamin D in the skin has no effect on the blood cholesterol level.

3.5.2 Comment

3.5.2.1 Does the skin's Vitamin-D Production Influence the Blood Cholesterol Level?

The amount of 7-DHC converted into previtamin D during Vitamin-D synthesis in the skin is too small to sustainably influence the concentration of cholesterol in the blood (Reichrath 2021; Wacker and Holick 2013). The resulting serum concentration of Vitamin D is too low for this.

▶ **Note** The formation of Vitamin D in the skin has no effect on the blood cholesterol level.

3.6 Misconception 6: *The Skin Produces Only a Few Hundred IU of Vitamin D Daily, Even Under Sufficient UV-B Influence*

3.6.1 Correction

According to studies, our skin can produce more than 20,000 IU of vitamin D on a sunny day.

3.6.2 Comment

3.6.2.1 How much Vitamin D Can Our Skin Produce?

The data on the daily synthesis capacity per cm^2 of skin vary between 9 and 40 ng, which, with an average body surface area of an adult of 1.73 m^2, corresponds to a total synthesis capacity between 156 and 692 µg (approx. 6000–28,000 IU) (Reichrath 2021; Wacker and Holick 2013). An important individual influencing factor of vitamin D synthesis is the skin type, which can be divided into 6 types according to a common classification (by Fitzpatrick, Fig. 3.4): skin type I (freckles, red hair and green eyes), skin type II (light skin, blonde hair, blue eyes), skin types III–IV (brown/dark hair, brown eyes, moderate to strong skin pigmentation), skin types V–VI (black hair, dark eyes, brown to black skin). The maximum synthesis capacity of vitamin D can be reached in light-skinned individuals depending on the intensity of UV radiation (UV index) after just a few minutes of whole-body exposure, while in dark-skinned individuals

3.6 Misconception ϵ: The Skin Produces ...

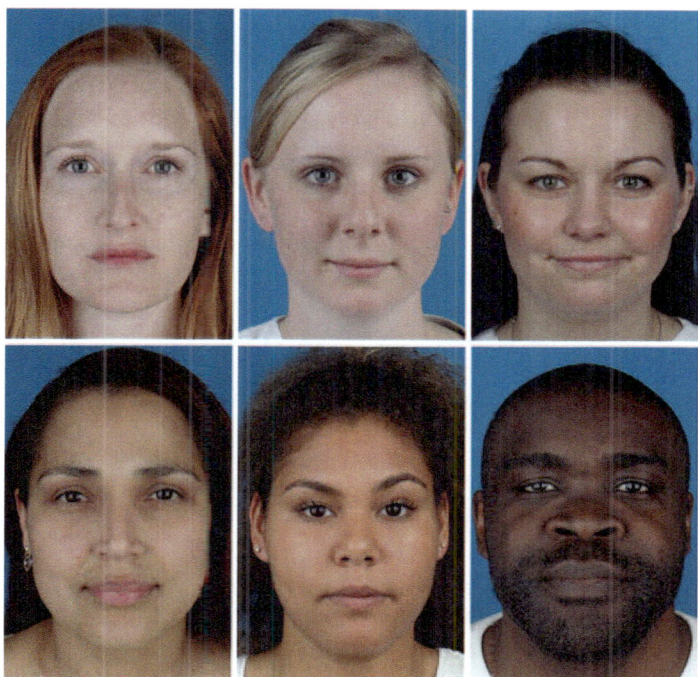

Fig. 3.4 The 6 skin types of humans according to Fitzpatrick (I–VI in ascending order from top left to bottom right). (*Source*: Singer et al. 2016)

it can take up to 2 hours. Traditionally, skin type V–VI is found near the equator (where there is high intensity of UV-B radiation all year round), skin type I, on the other hand, is found on the British Isles (geographical location relatively far north; known for rainy weather and low intensity of UV-B radiation).

Although various studies in northern regions have shown a pronounced association between skin color and vitamin D supply, this relationship is not undisputed. This is because vitamin D synthesis mainly takes place in the upper cell layers of the epidermis (stratum granulosum and stratum spinosum of the epidermis), while the pigment cells (melanocytes), which significantly determine the skin type, are located in deeper cell layers of the epidermis (stratum basale). Some experts believe that the reduced content of trans-urocanic acid in sweat and also on the skin of light-skinned individuals (alongside melanin one of the most important natural UV-absorbing molecules that protect the skin from UV radiation) results in more effective vitamin D synthesis. In general, the synthesis capacity of the skin for vitamin D is comparable in dark-skinned and light-skinned individuals. Only the exposure time to UV-B radiation or the dose must be higher in dark-skinned individuals in order to produce the same amounts of vitamin D.

Another factor influencing the skin's vitamin D synthesis is age. It has been shown that the 7-DHC content of the skin and consequently also the synthesis ability for vitamin D decrease by a factor of 3 in old age compared to a 20-year-old. UV-A rays (320–400 nm), which unlike UV-B rays also penetrate window glass, destroy vitamin D.

▶ **Note** According to studies, our skin can produce more than 20,000 IU of vitamin D in a day with sufficient UV-B exposure.

3.7 Misconception 7: *A Healthy Person Can Produce Too Much Vitamin D From Extensive Sunbathing and Thereby Endanger Health*

3.7.1 Clarification

In healthy people, even after extensive sunbathing, there is no risk of an oversupply of the biologically active Vitamin D metabolite $1,25(OH)_2D_3$. This is because as soon as more $1,25(OH)_2D_3$ is produced in the body than desired, its new formation is throttled (Reichrath 2021; Wacker and Holick 2013).

3.7.2 Comment

3.7.2.1 Can I "Poison" Myself with Vitamin D Through Too Intense Sunbathing?

No, because a Vitamin D intoxication, i.e., an excess of Vitamin D, is prevented by several mechanisms. With prolonged UV-B exposure of the skin, the synthesis of previtamin D_3 reaches a plateau. The amount of previtamin D_3 remains in balance with continued UV irradiation, as *previtamin D_3* is photolabile and is rapidly degraded by optical radiation. More specifically, with prolonged UV-B irradiation, previtamin D_3 is degraded to *Lumisterol* and *Tachysterol* within 3 hours before it can isomerize to Vitamin D_3. Until recently, these metabolic products were thought to be largely physiologically inactive. Even with strong sunlight, an excessive accumulation (cumulation) of Vitamin D_3 in the skin is prevented by the subsequent formation of these by-products. The Vitamin D_3 formed from previtamin D_3 is also rapidly degraded under the influence of optical radiation, i.e., it is photolabile. If Vitamin D_3 is not quickly transported out of the skin via the bloodstream, at least three other metabolites are formed from it under the continued influence of UV-B and UV-A radiation (up to 345 nm), which are currently believed to be biologically less active: *Suprasterol-1* and *-2* and *5,6-Transvitamin D_3*.

In addition, an excess of Vitamin D effect is also prevented by the fact that biologically active $1,25(OH)_2D_3$ is increasingly converted into biologically inactive metabolites (initially into $24,25(OH)_2D_3$ in the liver and other organs by the activation of a so-called P450 enzyme (CYP24A1, Vitamin D-24-Hydroxylase). Therefore, even the most intense sun exposure cannot cause Vitamin D intoxication (poisoning) in healthy people. A short sun exposure (with a sufficiently high UV-B component) for a few minutes produces a comparable amount of Vitamin D_3 as exposure over a longer period This protects the body from Vitamin D intoxication due to too intense UV radiation. For these reasons, even after extensive sunbathing, there is no risk of an oversupply of $1,25(OH)_2D_3$ in healthy people. In healthy individuals, there have been no reports of "poisoning" with 1,25(OH)$_2$D$_3$ even after extensive sunbathing.

▶ **Note** In healthy people, even after extensive sunbathing, there is no risk of "poisoning" with Vitamin D. This is because as soon as more biologically active 1,25(OH)$_2$D$_3$ is produced in the body than needed, its new formation is throttled via feedback mechanisms. In healthy individuals, there have been no reports of "poisoning" with 1,25(OH)$_2$D$_3$ even after extensive sunbathing.

3.8 Misconception 8: *UV-B Irradiation of Dairy Cows Has No Effect on the Vitamin D Content of Cow's Milk*

3.8.1 Correction

The vitamin D content of cow's milk depends on the conditions in which the dairy cows are kept. Just as in human skin, the UV rays of sunlight also cause the production of vitamin D in the fur of mammals.

3.8.2 Comment

3.8.2.1 Does the Vitamin D Content of Cow's Milk Change Due to Its UV-B Irradiation?

In the past, when dairy cows were still grazing in the pasture and grass was their main food, cow's milk was an important source of vitamin D, at least in the summer. However, this is no longer the rule. Because cow's milk now contains very little vitamin D, regardless of whether it is summer or winter. This development is also problematic because without vitamin D, the body can only utilize milk calcium to a limited extent. However, their vitamin D production can be increased all year round by a simple, inexpensive and ingenious method, namely the exposure of the cows to artificial or natural UV-B rays (Reichrath 2021; Wacker and Holick 2013). Under the rays of the sun or fluorescent tubes, the vitamin D concentration in the serum increases (Hymøller et al. 2017). With such special lamps, cows can produce 20 times more vitamin D_3 in the milk. This achieves values of 2.0 μg vitamin D per 100 g milk. According to press reports, about 10 years ago one of these fluorescent tubes, which mimic sunlight, cost about 190 €, with the total costs for installation in a stable with 70 cows being about 3500 €. So far, several farmers in Bavaria and Austria have installed the lighting system in the stables.

▶ **Note** UV-B irradiation of dairy cows increases the vitamin D content of cow's milk.

3.9 Misconception 9: *UV Irradiation of Food Has No Effect on Its Vitamin D Content*

3.9.1 Correction

This statement is not correct, as the vitamin D content of certain foods can be significantly increased by their UV irradiation.

3.9.2 Comment

3.9.2.1 Can the Vitamin D Content of Certain Foods Be Increased by Their UV-B Irradiation?

Yes. For example, mushrooms (including yeasts) contain the mycosterin ergosterol, which is converted into biologically active vitamin D_2 (ergocalciferol) by UV-B irradiation (Kalaras et al. 2012; Reichrath 2021; Wacker and Holick 2013). A study at the University Hospital Freiburg showed that cultivated mushrooms can produce significant amounts of vitamin D_2 (491 μg or 19,640 IU per 100 g cultivated mushrooms) by treatment with UV rays. The administration of these enriched cultivated mushrooms was equivalent to the intake of vitamin D_2 as a dietary supplement. Similar results were achieved with other types of mushrooms, including shiitake, maitake and shimeji. Values of up to 267,000 IU of vitamin D per 100 g of shiitake mushrooms were achieved through 14 hours of sun exposure.

> **Note** UV-B irradiation of certain foods increases their vitamin D content.

3.10 Misconception 10: *The Impact of Changes in Our Vitamin D Supply on Vitamin D Status is Difficult to Estimate*

3.10.1 Clarification

The change in vitamin D status can be well estimated both after intake of vitamin D and after UV exposure of the skin (Fig. 3.5).

3.10.2 Comment

3.10.2.1 How Does Vitamin D Status Change After Intake of Vitamin D Compared to After UV-B Exposure?

The minimal erythema dose (MED) is the lowest UV-B dose whose application leads to slight skin redness (erythema). It varies greatly from person to person. According to studies, sunbathing about 25% of the body surface with a UV-B dose of up to a third or half of this MED should have effects on the vitamin D status in the blood, which are comparable to the single intake of about 1500 to 2000 IU (250 μg to 500 μg) (Reichrath 2021; Wacker and Holick 2013). According to results from Cashman et al. (Cashman et al. 2008; Cashman 2020), the daily intake of 1000 IU (25 μg) vitamin D (i.e., about 365,000 IU vitamin D/year) achieves the target value of a 25(OH)D serum concentration of >20 ng/ml (50 nmol/l) in the winter months in about 95–97.5% of the Irish population. The above recommended sunbathing strategy (each about 2000 IU, 3×/week over 26 weeks) leads to the synthesis of vitamin D in the skin to an extent that

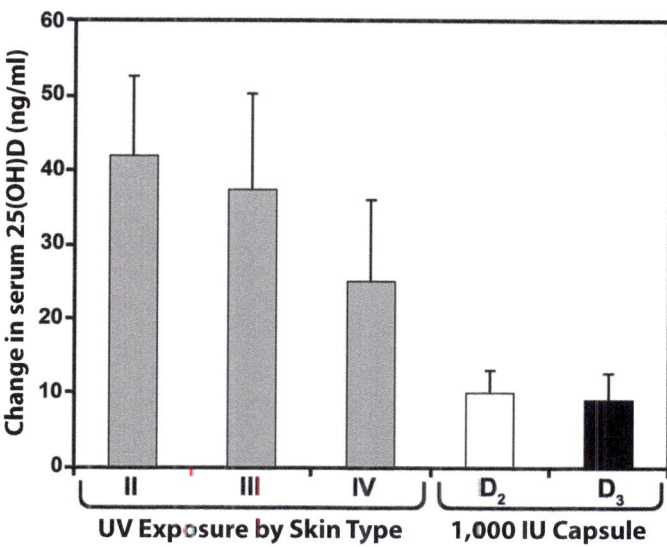

Fig. 3.5 Comparison of the increase in serum concentration of 25(OH)D after UV-B irradiation and after intake of vitamin D as a dietary supplement (supplement). Note the decrease in 25(OH)D serum concentration with increasing skin pigmentation ("Skin Type" II–IV). (Courtesy of Dr. Holick).

is only roughly equivalent to the oral intake of 156,000 IU/year. It is therefore unlikely that sufficient amounts of vitamin D will be formed through such sunbathing behavior. It must be taken into account that during the winter months (October–March) no vitamin D can be formed in the skin due to the too low intensity of natural UV-B radiation in our region at this time of year. Because in these months, in the *"vitamin D winter"*, the angle of incidence of sunlight in Central Europe is so flat that not enough UV-B radiation reaches the earth's surface through the atmosphere for sufficient vitamin D production.

▶ **Note** Changes in vitamin D status can be estimated both after intake of vitamin D and after UV exposure of the skin.

References

Cashman KD. Vitamin D deficiency: defining, prevalence, causes, and strategies of addressing. Calcif Tissue Int. 2020;106(1):14–29. https://doi.org/10.1007/s00223-019-00559-4. Epub 2019 May 8. PMID: 31069443.

Cashman KD, Hill TR, Lucey AJ, Taylor N, Seamans KM, Muldowney S, Fitzgerald AP, Flynn A, Barnes MS, Horigan G, Bonham MP, Duffy EM, Strain JJ, Wallace JM, Kiely M. Estimation of the dietary requirement for vitamin D in healthy adults. Am J Clin Nutr. 2008;88(6):1535–42. https://doi.org/10.3945/ajcn.2008.26594. PMID: 19064513.

Hymøller L, Jensen SK, Kaas P, Jakobsen J. Physiological limit of the daily endogenous cholecalciferol synthesis from UV light in cattle. J Anim Physiol Anim Nutr (Berl). 2017;101(2):215–21. https://doi.org/10.1111/jpn.12540. Epub 2016 Jul 16.PMID: 274212471.

Kalaras MD, Beelman RB, Holick MF, Elias RJ. Generation of potentially bioactive ergosterol-derived products following pulsed ultraviolet light exposure of mushrooms (Agaricus bisporus). Food Chem. 2012;135(2):396–401. https://doi.org/10.1016/j.foodchem.2012.04.132. Epub 2012 May 7. PMID: 22868105.

Krause R, Stange R. Lichttherapie. Berlin, Springer; 2012.

Reichrath J, Herausgeber. Sonne – die Dosis macht's! Berlin/Heidelberg: Springer; 2021.

Singer S, Schwarz T, Berneburg M. Phototherapie. Eine Einführung in die Wirkmechanismen und Anwendungsgebiete. Wiesbaden: Springer; 2016.

Wacker M, Holick MF. Sunlight and vitamin D: a global perspective for health. Dermatoendocrinol. 2013;5(1):51–108. https://doi.org/10.4161/derm.24494. PMID: 24494042; PMCID: PMC3897598.

Vitamin D, Skin Types, and Sunscreen

4

Contents

4.1	Misconception 1: *The Various Human Skin Types are a Whim of Evolution Without Biological Significance—And also Without Relevance for Cutaneous Vitamin D Production*	63
4.2	Misconception 2: *A Mild Sunburn is Harmless to Health*	65
4.3	Misconception 3: *The Use of Sunscreen Does Not Reduce the UV-Induced Formation of Vitamin D in the Skin*	68
4.4	Misconception 4: *The Use of Sunscreens Poses No Health Risk*...	70
4.5	Mistake 5: *The Use of Sunscreen Does Not Pose an Environmental Burden*	79
4.6	Mistake 6: *The Right Choice of Sunscreen Protects Against the Sun Without Reducing Vitamin D Synthesis*	82
References		83

4.1 Misconception 1: *The Various Human Skin Types are a Whim of Evolution Without Biological Significance—And also Without Relevance for Cutaneous Vitamin D Production*

4.1.1 Correction

Although the exact effects of the evolution of skin colors are currently controversially discussed among scientists, their general importance for human evolution and for our health is undisputed.

© The Author(s), under exclusive license to Springer-Verlag GmbH, DE, part of Springer Nature 2025
J. Reichrath, *Vitamin D in Focus*,
https://doi.org/10.1007/978-3-662-71341-9_4

4.1.2 Comment

4.1.2.1 The Evolution of Skin Types

Many scientists believe that the development of different skin types is also closely linked to human evolution. The evolution of skin colors and their significance is currently controversially discussed among scientists (Reichrath 2021; Wacker and Holick 2013). There are different hypotheses that show interesting correlations. I would like to introduce the most important ones to you here.

According to a widely accepted hypothesis, Europeans have lighter skin than Africans because they were exposed to fewer sun rays during the course of evolutionary history after settling in regions in the cold north. As a rule, people who live in warmer—thus sunnier—regions of the earth have darker skin than people who live in colder regions. Biologically, this phenomenon can be explained by Darwin's theory of evolution. According to this, living beings change from generation to generation through "random" gene mutations. However, those with positive effects are preferred for inheritance. Living beings whose gene mutations bring a positive effect thus have an advantage over others in the struggle for survival and reproduction. This is referred to as "survival of the fittest" (survival of the strongest). But what does this mean for the evolution of skin color? There are various hypotheses about the development of different skin types. Modern man—Homo sapiens—only appeared very late in the history of life. If one were to equate the history of the earth with a calendar year, he would only exist since December 31. It is now assumed that the first modern humans probably developed in equatorial, sun-rich regions of Africa. As an optimal adaptation to the local environmental conditions, they had a strongly pigmented skin, which enabled both: protection against the skin cancer-promoting effect of UV radiation and guarantee of sufficient vitamin D synthesis (Xiang et al. 2015). According to a theory favored by many scientists, the first humans had to first develop a light skin type in order to then be able to populate equator-distant and sun-poor regions like the northern hemisphere. When settling in sun-poor regions, a light skin type may then have represented a decisive selection advantage. Because only the development of new, less strongly pigmented skin types also guaranteed sufficient vitamin D production in sun-poor regions. It is known that dark skin is able to block the penetration of UV radiation up to a sun protection factor of about 8, which, however, due to the reduced radiation effect, can also reduce the skin's vitamin D production by up to 90%.

▶ **Note** A person's skin type affects their cutaneous vitamin D production.

4.2 Misconception 2: *A Mild Sunburn is Harmless to Health*

4.2.1 Correction

This assessment is wrong! The sunburn is not suitable as an early warning system to avoid skin cancer. Because the skin redness caused by sun radiation in the case of sunburn usually only becomes noticeable with a delay of one hour or more (Reichrath 2021; Wacker and Holick 2013). Especially if one has stayed in the sun during this period, the skin has already been irreversibly damaged. Because before sunburn occurs, the skin is already affected. Although this damage is not immediately visible, sunburn triggers processes that can later lead to a variety of consequential damage, from premature skin aging to skin cancer (Green et al. 1985).

4.2.2 Comment

4.2.2.1 How Long Can I Stay in the Sun Without Risking Sunburn?

An important question when dealing with the sun is: How long can one enjoy its radiation without burning the skin? Unfortunately, one usually only notices if sunbathing was too long when it is too late. This is because humans do not have a specific sense with which they can detect and measure UV radiation. Therefore, they are not able to sound the alarm in time when there is too much sun. Among other things, due to the lack of an early warning system, we often find it difficult to estimate how long sunbathing can last without risking sunburn. But there are also other reasons that make it difficult to estimate the duration one can stay in the sun without burning the skin. This is because the intensity of the sun's radiation that reaches our skin (and thus its impact) depends on many different conditions. Here, factors that determine the strength of near-ground UV radiation (including time of day/year, geographical location of the place of stay such as proximity to the equator, altitude above sea level), as well as factors that influence the individual sensitivity of our skin and our body to solar radiation (including skin type, extent of pre-tanning of the skin, intake of medications that influence the radiation sensitivity of the skin), play an important role. It is very difficult to estimate the influence of all these factors in a particular situation. To facilitate this, terms such as "*minimal erythemal dose*" (MED) have been defined and tools such as the so-called UV index have been developed to help us estimate the respective strength of UV radiation.

▶ **Note** Every sunburn endangers our health. Humans do not have an early warning system that could warn them of UV damage (sunburn).

4.2.2.2 Self-Protection Time, UV Index and the Shadow Rule

First, however, I would like to discuss another tool that is always and everywhere reliably applicable and constantly accompanies us in the sun. You probably already know what I mean: our shadow. When the sun is high in the sky at noon between April and September, the UV radiation is particularly intense. To assess the risk of sunburn, there is a simple and extremely useful rule of thumb. This is based on the universally valid basic rule that the smaller your own shadow, the higher the sun is in the sky. People with an average Central European skin type (skin type

4.2 Misconception 2: *A Mild Sunburn …*

II–III according to Fitzpatrick) have a low risk of burning for their unprotected skin when the shadow is significantly longer than the body length. If the shadow is about a third longer than the body length, the intensity of the UV radiation is in a range that is even low-risk for fair-skinned people. This situation occurs in Germany from the beginning of October to mid-March, even at noon.

An important advantage of this shadow rule is that the relationship between the length of the shadow and the UV intensity is almost independent of the time of year and day, as well as the geographical location. Thus, the sun in the tropics does not burn significantly stronger than in Central Europe—however, it is high in the sky all year round, so that UV indices of 7–10 are regularly reached at noon under clear skies. But what do we understand by the UV index and how can it help us deal with the sun? This estimate was developed by the World Health Organization (WHO) as a tool to assess the strength of solar radiation. The UV index is uniform worldwide. It is determined daily on an open-ended scale and indicates the expected daily peak value of UV irradiance at ground level. As the UV index increases, so does the risk of sunburn. To determine the ground-level UV radiation and the UV index, the Federal Office for Radiation Protection (BfS), the Federal Environment Agency (UBA), the German Weather Service (DWD) and other institutions operate a nationwide measurement network with a total of ten measuring stations. The daily measured values of the strength of the sunburn-effective ground-level UV radiation at these measuring stations are transmitted to the measurement network center in BfS Munich-Neuherberg, where they are evaluated and made available to the public on the Internet in the form of UV index values. Based on this measurement network data and satellite data from the German Weather Service (DWD), the BfS and other institutions publish forecasts of expected daily peak values of the UV index for ten regions in Germany on the Internet (www.bfs.de/uv-prognosis). Thus, the UV index is usually well above 3 at noon in summer. When the shadow is the same size as your own body, the UV index is about 4.

Since the forecasts for the UV index are determined for larger regions, in which the ground-level UV radiation can vary significantly at different locations, they only provide inaccurate estimates of the possible UV exposure. Since these estimates do not represent precise measurements of the actual UV exposure of an individual person, they cannot be used to calculate individual self-protection times. The self-protection time refers to the maximum duration that one can expose the skin to the sun in the course of a day without it becoming red. Depending on the skin type, the self-protection time can vary in duration between 3 min for very fair skin (skin type I) and 40 min for Mediterranean brown skin (skin type IV). It is standardized at UV index 8 (midday sun in summer in Central Europe). With a higher UV index (high mountains, Mediterranean, tropics) and in a reflective environment (water, snow, sand), the self-protection time is significantly lower. On the other hand, it is higher for skin that has been accustomed to the sun and tanned over several weeks (often not after using self-tanners). All stays in the sun within the last 24 hours must be taken into account.

▶ **Note** An important measure for assessing the individual sensitivity of our skin to solar radiation is the so-called minimal erythema-effective dose (MED; unit: joules per square meter, J/m^2). The MED refers to the smallest dose of UV radiation that leads to the development of skin redness. On a sunny day in Central Europe at noon with a UV index (UVI) of 8, people with skin type II can theoretically reach this dose after about 20 minutes.

4.3 Misconception 3: *The Use of Sunscreen Does Not Reduce the UV-Induced Formation of Vitamin D in the Skin*

4.3.1 Correction

For the formation of vitamin D, it is absolutely necessary that UV-B rays reach the skin (Reichrath 2021; Wacker and Holick 2013). Without the influence of UV-B rays, our skin cells cannot produce vitamin D. The skin's hormone factory then lacks, so to speak, the energy required to maintain production. Therefore, it is easily understandable that the use of sunscreen reduces the formation of vitamin D in the skin. This is because sunscreen prevents a large part of the UV-B rays needed for the formation of vitamin D from reaching the skin cells.

4.3.2 Comment

4.3.2.1 Does the Use of Sunscreen Affect Cutaneous Vitamin D Synthesis?

A multitude of experimental and clinical studies show that the consistent use of sunscreen affects a person's vitamin D status, i.e., reduces their 25(OH)D concentration in the blood (Reichrath 2021). Here is a brief summary of such a study for better understanding (Querings et al. 2006). To prevent rejection reactions, patients usually have to take drugs that suppress the immune system after organ transplantation. Many of these so-called immunosuppressants, however, increase the risk of skin cancer caused by UV radiation. Therefore, these patients must maintain consistent sun protection, which also relies on the use of sunscreen. In several studies, this patient group was found to have a lower vitamin D status (compared to the control groups). In populations that practice consistent sun protection and therefore have a high risk of vitamin D deficiency, the vitamin D level should be regularly checked and supplementation started if necessary.

However, the study situation regarding the question of whether the use of sunscreen under all "real-life" conditions increases the risk of vitamin D deficiency is not uniform. In some studies, the use of sunscreen was not associated with a reduced vitamin D status. This is explained by the fact that the sunscreen was not properly applied in such cases (too little cream applied, not all skin areas to be protected were creamed, etc.), possibly because the motivation for consistent sunscreen application was lower than in the studies outlined above on patients after organ transplantation. Another reason could be that the characteristic "sunscreen use" queried in so-called case-control studies merely represents a "marker" for so-called "sun worshippers". The study participants could therefore have spent significantly longer time in the sun after applying sunscreen than the subjects who did not use sunscreen. In addition, it has been shown that certain UV filters disrupt the vitamin D synthesis of the skin as "endocrine disruptors" (Abdi et al. 2022).

> **Note** The use of sunscreen reduces the formation of vitamin D in the skin.

4.4 Misconception 4: *The Use of Sunscreens Poses No Health Risk*

4.4.1 Correction

This statement is false. Because certain UV filter substances

- do not just stay on the skin, but also enter our body and the environment,
- are transferred to the unborn child during pregnancy and also enter breast milk,
- are now detectable in all surface waters (from streams to oceans) and in the food chain (including fish, other marine creatures),
- have well-documented possible side effects such as photo-toxic/photoallergic skin reactions,
- may possibly transform into carcinogenic substances after the sunscreen's expiration date,
- can disrupt our hormone metabolism as *"endocrine disruptors"*.

4.4.2 Comment

4.4.2.1 Do UV Filters Stay on the Skin or Do They Also Enter the Body?

Is there anything against using sunscreen? Unfortunately, yes, because recently many of the organic UV filter substances contained in sunscreens are increasingly making headlines (Abdi et al. 2022; Ferraris et al. 2020; Inderbinen et al. 2022; Murawski et al. 2021; Reichrath 2021; Schlumpf et al. 2010; Siller et al. 2019; Wnuk et al. 2022). Several studies have convincingly shown that these important ingredients of sunscreens are absorbed through the skin and enter the body. In a large US study (National Health and Nutrition Examination Survey), the chemical UV protection filter benzophenone-3 was detected in urine samples of 96.8% of over 2500 examined children (from 6 years) and adults. In another study conducted by the US Food and Drug Administration (FDA), it was also shown that four UV filters contained in commercially available sunscreens (avobenzone, oxybenzone, octocrylene, and ecamsule) passed into the blood and were detectable there. As part of this study, 24 participants had to apply the recommended amount of sunscreen (2 mg/cm^2) to 75% of their body surface four times a day for a treatment period of 4 days. Blood analyses showed that all four of these commercially available organic UV filters were detectable in the blood after the first day. Since they are subsequently filtered out of the blood by the kidneys and excreted with the urine, they also enter the environment.

Other studies have shown that certain chemical UV filters (including homosalate, methylbenzylidene camphor, benzylidene camphor) are transferred to the unborn child via the placenta during pregnancy and, when used by breastfeeding mothers, enter breast milk in a very high percentage (up to 82.5%). As a result, they are ingested by infants. This has been demonstrated for two widely used organic UV filters (octocrylene, methylbenzylidene camphor), which are also contained in products advertised as "*sunscreen for children*".

4.4.2.2 Does the Use of Sunscreen Pose Health Risks?

Why is the detection of chemical UV filters after the application of sunscreen in the body, as well as in breast milk and placenta, significant? Although the results of these studies do not allow conclusions about possible harmful health effects of sunscreens, they do raise many questions about the safety of these UV filters.

Currently, a serious side effect that can occur when using organic UV filters is being intensively discussed: the causing of hormonal side effects. It has been found that organic UV filter substances can disrupt the hormonal balance. This phenomenon is referred to in the English literature as *"endocrine disruption"*. Such effects are based on a strong binding of these hormonally active substances to certain corresponding hormone receptors in target cells. This binding activates the receptors, which in turn directly regulate numerous genes. Of particular interest is an estrogen-like activity. This refers to an effect similar to that of the female sex hormone estradiol, which is mainly produced in the ovaries in humans. UV filter substances can also manipulate other hormone systems, including thyroid hormones in addition to other sex hormones.

In addition to the hormonal effects of some organic UV filter substances, sunscreens have many other possible unwanted side effects. According to the Federal Institute for Risk Assessment (BfR; www.bfr.bund.de), certain sunscreens can be hazardous to health after their expiration date, as their UV filters may possibly transform into carcinogenic substances after too long storage (including octocrylene and benzophenone). It has also been known for many years that organic UV filters can trigger allergic skin reactions. These often clinically manifest as sunburn-like itchy redness in the areas of skin treated with sunscreen. They can also occur as so-called scattered foci in their surroundings not treated with sunscreen. Often, the simultaneous exposure to sunlight is required for such reactions to occur. These reactions are then referred to as photoallergic.

Clinically similar to these allergic skin reactions are the so-called phototoxic reactions, which can also occur as unwanted side effects when using organic UV filter substances. These often clinically manifest as sunburn-like inflammations in the treated skin areas, similar to the photoallergic reactions. However, phototoxic reactions usually do not form scattered foci. They are also not caused by an allergic reaction. Instead, the causing substance, in this case the organic UV filter, causes a so-called photosensitization. This refers to an intensification of the damaging effect of sunlight on skin cells, which does not have to be caused by allergic mechanisms.

The so-called Mallorca acne (Acne aestivalis) can also occur as an unwanted side effect after the application of sunscreen. It often affects young people and manifests a few hours after sunbathing as itchy nodules in sun-exposed skin areas. This is usually the case after contact with the first stronger sun rays in spring. The cause of Mallorca acne is today considered to be a combination of the affected person's genetic predisposition, the use of unsuitable care or sun protection products, and the effect of UV-A radiation. These factors lead to the formation of so-called free radicals, which can damage skin cells and trigger inflammation with the involvement of an autoimmune reaction. Mallorca acne is thus not a variant of classic acne. The use of modern sun protection products with reduced lipid content and high UV-A protection has made Mallorca acne less common than a few years ago.

4.4.2.3 How are Sunscreens Approved?

Overall, many scientists currently see a great need to more comprehensively investigate the safety of UV filters. The approval processes for cosmetics, and thus also for sunscreens, are significantly less demanding than those for pharmaceuticals. They are monitored in Germany by the Federal Institute for Risk Assessment (BfR) and in Europe by the Scientific Committee on Consumer Safety (SCCS). Although UV filters, like preservatives, dyes, and many other ingredients and additives, must undergo an approval process in which the manufacturers must guarantee the safety of their products through safety assessments. This usually involves using cell culture models to conduct various studies on potential harmful effects. For example, in suitable cell culture models (organ cultures of mammary gland

and uterine tissue), hormonal effects were unequivocally shown for most of the UV filter substances examined (benzophenone-3, homosalate, 4-methyl-benzylidene camphor, octyl methoxycinnamate, octyl dimethyl PABA, octocrylene). These were comparable to the effect of the female sex hormone estrogen. Many of these UV filters are very widely used. For example, octocrylene was found in 12 of the 23 sunscreens examined by *Öko-Test* in 2015.

In addition to investigating a possible hormonal effect, many further analyses can be carried out in cell culture experiments. For example, one can investigate whether the growth behavior or the division speed of the cells change after the addition of the UV filters. Experimental studies have shown an acceleration for several organic UV filters. However, these cell culture experiments cannot be directly transferred to the situation in humans: they cannot prove a hormone-like or other effect of the UV filters after sunscreen application in humans. There are many reasons for this. One of them is the fact that not everything that is rubbed onto the skin actually ends up inside the body. Experiments on pig or human skin have shown that this proportion is less than 1% for many organic UV filter substances. More meaningful studies, such as so-called pharmacokinetic studies or studies on drug absorption through the skin, are generally not available for sunscreens.

Animal testing is generally avoided today. Since 2013, it has been completely forbidden in the European Union to test cosmetic ingredients on animals. Nevertheless, the safety of many of the UV filter substances used in sunscreens today was still tested on animals. This is because these UV filters were approved beforehand. In such animal tests, for example, rats were given daily doses of UV filter substances in different dosages to observe possible changes in hormone levels or their reproductive ability. In other experiments, scientists examined rodents to see whether UV filter substances could stimulate the uterus to grow. This would indicate a hormone-like effect. In fact, as early as 2001, scientists at the University of Zurich were able to demonstrate increased uterine growth in rats after adding three different UV filter substances to their food. However, these results can also only be compared to a limited extent with the situation in humans after applying sunscreen. Because it makes a big difference whether these substances are ingested with food or through the skin. Both applications result in different amounts of absorption into the body. For this reason alone, the results obtained in animal experiments with rats fed UV filter substances cannot be directly transferred to humans. The European Union (EU) approval process for cosmetics does not provide for human trials.

In matters of consumer protection, the Scientific Committee on Consumer Safety (SCCS) advises the political decision-makers of the EU Parliament. To assess the safety of organic UV filters in humans, these experts base their assessments on the amount of UV filters in the entire sunscreen, their prescribed application amount, and their absorption rate through the skin. They first estimate the internally absorbed dose in humans. If results of unwanted effects are available from older animal experiments, the following rule applies: This estimated possible internally absorbed dose for humans must be at least a hundred times lower than the lowest dose that still showed unwanted effects in the animal experiment.

According to the Federal Institute for Risk Assessment (BfR), there is so far no evidence that sunscreen is carcinogenic. However, sunscreens with the UV filter octocrylene, whose expiration date has passed, could be hazardous to health, as this ingredient is said to transform into the potentially carcinogenic benzophenone over time, according to researchers. The current assessments of the SCCS do not indicate any hormone-like or other harmful effects of the organic UV filter substances approved in the EU in humans. However, in the opinion of many scientists, the data situation for some of the UV filters commonly used today is still insufficient.

4.4.2.4 Benefits and Risks of Sunscreens: What to Consider When Choosing?

When choosing a sunscreen, it is important to follow certain basic rules. Sunscreens should only be used because of their potential environmental (e.g., coral death) and health-damaging effects (e.g., acting as "endocrine disruptors") when better suited sun protection measures (e.g., avoiding the sun, seeking shade, wearing suitable clothing) are not possible or not desired for urgent reasons. It is important to note that not all clothing provides sufficient protection against UV rays. For UV protection, special UV protective clothing can be used, which is offered by numerous providers for various activities (leisure, vacation, work) and blocks up to more than 99% of UV radiation. Only when avoiding sunlight or protecting against it with suitable clothing is not possible or desired, should the use of sunscreens be considered.

If you decide to use a sunscreen, the use of mineral (physical) UV filters is recommended. However, these should not be based on nano-technology (see below). Organic (chemical) UV filters should also be avoided if possible, in my opinion, due to their potential hormonal and other unwanted side effects and the environmental impact associated with their use. Unfortunately, many products often combine several mineral and chemical UV filter substances. This is done to cover as broad a range as possible of the wide spectrum of sunlight and to achieve the high sun protection factor (SPF) often demanded by doctors and pharmacists. However, an SPF of 50 is usually only useful for certain rare diseases that cause increased light sensitivity, including albinism, xeroderma pigmentosum, porphyrias. Because an SPF of 30 blocks only about 0.9% less UV radiation than an SPF of 50 and is therefore usually completely sufficient. To achieve effective UV protection, it is much more important to apply sunscreens consistently, i.e., to generously rub all body areas exposed to the sun with sunscreen (see below).

4.4.2.5 Sunscreen: Caution with Infants, Toddlers, Nursing Mothers, and Pregnant Women

When using sunscreens, special attention should be paid to infants, toddlers, nursing mothers, and pregnant women. In my opinion, organic UV filters should not be used in this group of people due to the possible hormonal (estrogen-like) and other side effects, until comprehensive clinical studies have shown their safety. If one decides to use sunscreens for these individuals, preferably mineral (physical) UV filters should be used. Especially products that are in the classic "*micronized*" form (which often leaves a white layer on the skin after application)—even though these can also not be considered completely harmless in all cases. Sunscreens containing nanomolecular ingredients are particularly controversial. In these, the particle size of the mineral UV filters (often zinc or titanium dioxide) is less than 100 nm. This improves the spreadability of the sunscreens and mitigates the visually disturbing effect of the "white film" on the skin. It may also increase the effectiveness of UV protection. However, these obvious advantages are also associated with risks. It has been proven that nanomolecular substances are capable of acting as carrier substances ("*nanocarrier*") to transport proteins through the skin into the body. This could potentially cause unwanted effects there.

Currently, mineral UV filters such as titanium dioxide and zinc oxide (ZnO) are also considered safe as nanoparticles in sunscreens and other cosmetics up to a concentration of 25%. This is because they generally do not penetrate deeper into the skin than the top layer of the epidermis (stratum corneum). This is mainly due to the fact that the molecules of the mineral UV filters, despite their nano-labeling, are 10 to 100 times larger than those of modern organic filtering substances. This makes it impossible for them to migrate through the small gaps between skin cells into deeper layers. In the EU, nanomolecular ingredients in sunscreens and other cosmetics have been subject to declaration since June 2013.

Sunscreen should be used with particular caution in children. This is because they have an increased risk of unwanted absorption of substances through the skin. Relative to body weight, the surface area of the skin in children is up to 3 times larger than in adults. This results in the amount of substances absorbed through the skin per kg of body weight being significantly higher in children than in adults. Another reason is the much higher number of sebaceous glands and hair follicles per cm^2 of skin surface in children compared to adults. These structures, which decrease in number with increasing age, have a funnel effect and thus represent "*entry gates*" for substances applied to the skin.

Even when using mineral (physical) UV filters, however, many questions remain unanswered. These concern not only physical and chemical aspects of zinc and titanium dioxide compounds, but also their application route and the condition of the skin as important additional influencing factors. Since inhalation into the lungs allows for much easier absorption into the body compared to absorption through the skin, nanomolecular particles in sun sprays are prohibited in Germany.

When choosing a sunscreen with as few burdened UV filters as possible, I recommend using product check apps like "ToxFox" from the German Federation for Environment and Nature Conservation (BUND). This allows the barcode of cosmetics and other products to be scanned with the mobile phone, providing information on the safety of the preparations.

4.4.2.6 Sunscreen: How to Apply?

When deciding to use sun protection products, the following points should be considered: First and foremost, physical sun protection products should be used for sun protection, if possible, in addition to avoiding exposure and using textiles. Skin areas that are not covered by clothing (often head, face, hands, arms, legs) should be carefully protected by sunscreen. It is important that the sun protection product is applied evenly and thickly to the affected body parts before sun exposure. The often suggested rule here is: "more is better". As a guide, one can assume that an adult needs about 30–40 ml of sunscreen for a single application to the entire body. It is also important to reapply the sunscreen regularly depending on the situation, for example after each stay in the water or when sweating heavily.

Depending on the intensity of the sun's rays, the duration of stay in the sun and the individual skin condition, a suitable sun protection factor should be chosen. It is important to note that the absolute sun protection time cannot be extended by regular reapplication, especially after bathing or heavy sweating. It should also be noted that the respective sun protection factor only refers to the protective effect against UV-A/B radiation. Whether and how effectively a particular sunscreen also protects against other spectral ranges of solar radiation (e.g., infrared rays or visible light) is often not investigated and is usually not indicated on the respective products. For this reason, one should avoid extending the stay in the sun due to the application of sunscreen. This increases the risk of potential health damage due to UV-independent effects of solar radiation on our body.

In strong sunlight, suitable sunglasses should be worn. However, even when wearing sunglasses, one should never look directly into the sun high in the sky (for example in summer at noon or on a beach holiday near the equator).

Purely plant-based products with wheat germ, raspberry seed or coconut oil are sometimes heavily advertised as a "*natural alternative*". However, their application usually does not provide sufficient sun protection.

▶ **Note** The use of certain sunscreens can endanger our health.

4.5 Mistake 5: *The Use of Sunscreen Does Not Pose an Environmental Burden*

4.5.1 Correction

The use of sunscreen burdens the environment (Conway et al. 2021; Fivenson et al. 2020; Reichrath 2021). Various UV filter substances, including octinoxate and oxybenzone, are held responsible for causing coral bleaching and damaging the genetic material of fish and corals.

4.5.2 Comment

4.5.2.1 UV Filter Substances in Surface Waters and in the Food Chain

The enormous environmental burden caused by UV filter substances has become increasingly clear in recent years (Conway et al. 2021; Fivenson et al. 2020; Schlumpf et al. 2010). The widespread use of sunscreens has serious impacts on our environment. By now, not only drinking water and surface waters including the seas, but also the food chains based on them are contaminated with these substances in all regions of our planet. In numerous regions, including Switzerland, Brazil, Canary Islands (Gran Canaria), and the Virgin Islands, UV filters have been detected in drinking water, rivers, and the sea. These substances are unfortunately not eliminated in the process of drinking water treatment. Overall, a large number of different substances have been detected in different waters (including homosalate, butyl methoxydibenzoylmethane, 4-methylbenzylidene camphor, diethylamino hydroxybenzoyl hexyl benzoate). The highest concentrations were found for benzophenone-3 (up to 6073 ng/l), octocrylene, and ethylhexyl methoxycinnamate. These substances are now also detectable in the food chain. For example, high concentrations of the UV filters octocrylene (up to 2400 ng/g) and 4-methylbenzylidene camphor (up to 1800 ng/g) were found in trout from small Swiss rivers into which wastewater was discharged.

4.5.2.2 UV Filter Substances, Coral Death, and Marine Life

Various UV filter substances, including octinoxate and oxybenzone, are accused of triggering coral bleaching and damaging the genetic material of fish and corals (Fig. 4.1). However, it should be noted that many of these thought-provoking findings come from tests in aquariums. Therefore, further investigations are absolutely necessary to further clarify the exact relationships. It is estimated that between 6000 and 140,000 tons of sunscreen are washed into the sea by swimmers and divers in the vicinity of corals every year. Four widely used UV filters (oxybenzone, butylparaben, octinoxate, and 4-methylbenzylidene camphor, all allowed in Europe, Canada; not in USA, Japan) have the ability to trigger coral bleaching even at extremely low dosages (one drop is enough to contaminate the 6.5-fold water volume of an Olympic swimming pool). In Hawaii (from 2021) and many other regions, such sunscreens are or will be banned. Toxic

4.5 Mistake 5: *The Use of Sunscreen ...*

Fig. 4.1 Sunscreen and coral death. Various UV filter substances, including octinoxate and oxybenzone, are suspected of triggering coral bleaching and damaging the genetic material of fish and corals. Strong contrast: (**a**) Vibrant, colorful corals as an expression of an intact environment. (**b**) Coral reef destroyed by coral bleaching in the Maldives. (*Source*: ©damedias/stock.adobe.com/225599023 (**a**); ©helivideo/stock.adobe.com/354200683 (**b**))

(poisonous) effects on marine organisms have also been shown for mineral (nano-ZnO) UV filters.

▶ **Note** The use of sunscreens burdens our environment.

4.6 Mistake 6: *The Right Choice of Sunscreen Protects Against the Sun Without Reducing Vitamin D Synthesis*

4.6.1 Correction

Unfortunately, it is not possible to fully protect oneself from the negative effects of sun rays by choosing the "right" sunscreen and at the same time ensure sufficient vitamin D synthesis (Reichrath 2021).

4.6.2 Comment

4.6.2.1 Theoretical Considerations and Experimental Investigations

It is not possible to develop a sunscreen that protects against the negative effects of sun rays and at the same time ensures sufficient vitamin D synthesis by "filtering out" certain wavelengths (Kockott et al. 2016). This is because the spectral range of UV-B radiation that causes skin cancer is largely identical to that which triggers vitamin D synthesis (Wacker and Holick 2013). Without the impact of UV-B rays, our skin cells cannot produce vitamin D. The hormone factory skin then lacks the energy required to maintain production. The intensity of the incident UV-B radiation is the main limiting factor for the skin's vitamin D production. Therefore, it is easy to understand that with every UV-B photon that is prevented from reaching the skin by the application of sunscreen, the formation of vitamin D in the skin is reduced. It is well documented in the scientific literature that sunscreen, depending on its respective sun protection factor, prevents a large part of the UV-B rays from reaching the skin cells. Thus, a large number of studies also show that the consistent use of sunscreen reduces a person's vitamin D status, i.e., their 25-hydroxyvitamin D concentration in the blood. Here is a brief example: To prevent rejection reactions, patients usually have to take drugs that suppress the immune system after organ transplantation. Many of these so-called immunosuppressants, however, increase the risk of skin cancer caused by UV radiation. Therefore, these patients must follow consistent UV/sun protection, which also relies on the use of sunscreen. In several studies, this patient group was found to have a reduced vitamin D status (compared to the control groups) (Reichrath 2021).

4.6.2.2 Further Clinical Investigations and Application Observations

However, the study situation regarding the question of whether intensive sun protection increases the risk of vitamin D deficiency is not uniform. In some clinical studies, the use of sunscreen was not associated with a reduced vitamin D status. This is explained by the fact that the sunscreen was not applied correctly (too little cream applied, not all skin areas to be protected were creamed, etc.) and that people stay in the sun longer after applying sunscreen than without sunscreen.

▶ **Note** Even with the right choice, a sunscreen cannot protect against the negative effects of sun rays without inhibiting vitamin D synthesis in the skin.

References

- Abdi SAH, Ali A, Sayed SF, Nagarajan S, Abutahir AP, Ali A. Sunscreen ingredient octocrylene's potency to disrupt vitamin D synthesis. Int J Mol Sci. 2022;23(17):10154. https://doi.org/10.3390/ijms231710154. PMID: 36077552; PMCID: PMC9456232.
- Conway AJ, Gonsior M, Clark C, Heyes A, Mitchelmore CL. Acute toxicity of the UV filter oxybenzone to the coral Galaxea fascicularis. Sci Total Environ. 2021;20(796):148666. https://doi.org/10.1016/j.scitotenv.2021.148666. Epub 2021 Jun 25. PMID: 34273823.
- Ferraris FK, Garcia EB, Chaves ADS, de Brito TM, Doro LH, Félix da Silva NM, Alves AS, Pádua TA, Henriques MDGMO, Cardoso Machado TS, Amendoeira FC. Exposure to the UV filter octyl methoxy cinnamate in the postnatal period induces thyroid dysregulation and perturbs the immune system of mice. Front Endocrinol (Lausanne). 2020;31(10):943. https://doi.org/10.3389/fendo.2019.00943. PMID: 32082254; PMCID: PMC7005579.

Fivenson D, Sabzevari N, Qiblawi S, Blitz J, Norton BB, Norton SA. Sunscreens: UV filters to protect us: Part 2-increasing awareness of UV filters and their potential toxicities to us and our environment. Int J Womens Dermatol. 2020;7(1):45–69. https://doi.org/10.1016/j.ijwd.2020.08.008. PMID: 33537395; PMCID: PMC7838327.

Green A, Siskind V, Bain C, Alexander J. Sunburn and malignant melanoma. Br J Cancer. 1985;51(3):393–7. https://doi.org/10.1038/bjc.1985.53. PMID: 3970815; PMCID: PMC1976961.

Inderbinen SG, Kley M, Zogg M, Sellner M, Fischer A, Kędzierski J, Boudon S, Jetten AM, Smieško M, Odermatt A. Activation of retinoic acid-related orphan receptor γ(t) by parabens and benzophenone UV-filters. Toxicology. 2022;23(471):153159. https://doi.org/10.1016/j.tox.2022.153159. Epub ahead of print. PMID: 35337918.

Kockott D, Herzog B, Reichrath J, Keane K, Holick MF. New approach to develop optimized sunscreens that enable cutaneous vitamin D formation with minimal erythema risk. PLoS One. 2016;11(1):e0145509. https://doi.org/10.1371/journal.pone.0145509. PMID: 26824688; PMCID: PMC4732611.

Murawski A, Schmied-Tobies MIH, Rucic E, Schmidtkunz C, Küpper K, Leng G, Eckert E, Kuhlmann L, Göen T, Daniels A, Schwedler G, Kolossa-Gehring M. Metabolites of 4-methylbenzylidene camphor (4-MBC), butylated hydroxytoluene (BHT), and tris(2-ethylhexyl) trimellitate (TOTM) in urine of children and adolescents in Germany – human biomonitoring results of the German Environmental Survey GerES V (2014–2017). Environ Res. 2021;192:110345. https://doi.org/10.1016/j.envres.2020.110345. Epub 2020 Oct 20. PMID: 33096061.

Querings K, Girndt M, Geisel J, Georg T, Tilgen W, Reichrath J. 25-hydroxyvitamin D deficiency in renal transplant recipients. J Clin Endocrinol Metab. 2006;91(2):526–9. https://doi.org/10.1210/jc.2005-0547. Epub 2005 Nov 22. PMID: 16303843.

Reichrath J, Herausgeber. Sonne – die Dosis macht's! Berlin/Heidelberg: Springer; 2021.

Schlumpf M, Reichrath J, Lehmann B, Sigmundsdottir H, Feldmeyer L, Hofbauer GF, Lichtensteiger W. Fundamental questions to sun protection: A continuous education symposium on vitamin D, immune system and sun protection at the University of Zürich. Dermatoendocrinol. 2010;2(1):19–25. https://doi.org/10.4161/derm.2.1.12016. PMID: 21547144; PMCID: PMC3084961.

Siller A, Blaszak SC, Lazar M, Olasz HE. Update about the effects of the sunscreen ingredients oxybenzone and octinoxate on humans and the environment. Plast Surg Nurs. 2019;39(4):157–60. https://doi.org/10.1097/PSN.0000000000000288. PMID: 31790045.

Wacker M, Holick MF. Sunlight and vitamin D: a global perspective for health. Dermatoendocrinol. 2013;5(1):51–108. https://doi.org/10.4161/derm.24494. PMID: 24494042; PMCID: PMC3897598.

Wnuk W, Michalska K, Krupa A, Pawlak K. Benzophenone-3, a chemical UV-filter in cosmetics: is it really safe for children and pregnant women? Postepy Dermatol Alergol. 2022;39(1):26–33. https://doi.org/10.5114/ada.2022.113617. Epub 2021 Feb 28. PMID 35369611; PMCID: PMC8953895.

Xiang F, Lucas R, de Gruijl F, Norval M. A systematic review of the influence of skin pigmentation on changes in the concentrations of vitamin D and 25-hydroxyvitamin D in plasma/serum following experimental UV irradiation. Photochem Photobiol Sci. 2015;14(12):2138–46. https://doi.org/10.1039/c5pp00168d. PMID: 26548800.

How Does Vitamin D Work? 5

Contents

5.1 Misconception 1: *The Vitamin D Formed in the Skin or Ingested with Food is Biologically Active* 87
5.2 Misconception 2: *1,25-dihydroxyvitamin D (1,25(OH)$_2$D) is the Only Biologically Active Vitamin D Metabolite* 88
5.3 Mistake 3: *All Effects of Vitamin D Formed in the Skin Can Be Compensated by Oral Intake of Vitamin D* 91
References 92

5.1 Misconception 1: *The Vitamin D Formed in the Skin or Ingested with Food is Biologically Active*

5.1.1 Clarification

Vitamin D is largely inactive. In order for it to exert its biological effects, it must first be converted into active substances in our body through chemical reactions.

5.1.2 Comment

5.1.2.1 Vitamin D: A Prohormone

Strictly speaking, vitamin D is a so-called "prohormone" (hormone precursor), as it is largely biologically inactive and only becomes biologically active through subsequent chemical reactions (so-called hydroxylations mediated by enzymes), such as in the liver (through the enzymes CYP27A1 and CYP2R1 present there) and kidney (via CYP27B1) into 1,25-dihydroxyvitamin D ($1,25(OH)_2D_3$, calcitriol), the classic, biologically active vitamin D hormone (Holick et al. 1980; Reichrath 2021; Wacker and Holick 2013).

5.2 Misconception 2: 1,25-dihydroxyvitamin D ($1,25(OH)_2D$) is the Only Biologically Active Vitamin D Metabolite

5.2.1 Clarification

This claim is incorrect, as in recent years, in addition to $1,25(OH)_2D_3$, numerous other biologically active vitamin D metabolites have been identified, the exact significance of which in many cases is still largely unknown.

5.2.2 Comment

5.2.2.1 Vitamin D-Mediated Gene Regulation: Endocrine and Autocrine/Paracrine $1,25(OH)_2D_3$ Activities and New Signaling Pathways

How is it possible for the vitamin D hormone system to perform so many different tasks in almost all organs of our body? How can it exert such a wide range of different effects in a temporally and spatially ordered manner? One reason is certainly the complex and widely branched structure of vitamin D metabolism.

5.2 Misconception 2: *1,25-dihydroxyvitamin D (1,25(OH)* …

This allows it to be regulated and fine-tuned at many different control points. The interconnections enable our body to adapt to the most diverse required physiological and pathophysiological states. The fine-tuning is mainly achieved through the formation of different vitamin D metabolites and corresponding receptors in target cells.

The formation of active vitamin D metabolites is usually mediated by specific cytochrome P450 enzymes and proceeds in several stages (Fig. 1.5). The enzymes thus determine which vitamin D metabolites are present at which time in which quantity and where they exert their effects. In addition to a non-genomic activity, which is predominantly mediated in target cells by rapid calcium influx, $1,25(OH)_2D_3$ and other biologically active vitamin D metabolites exert a significant part of their effect through the regulation of target genes. In this process, vitamin D metabolites first bind to corresponding receptors in the cell nucleus (Fig. 1.5). This activates these receptors and enables them to bind to specific DNA sections in target genes ("response elements"). This process turns genes on or off. It is currently assumed that $1,25(OH)_2D_3$ alone regulates more than 1000 genes via the activation of the classic vitamin D receptor (VDR).

$1,25(OH)_2D_3$ is not the only biologically active vitamin D metabolite, there are many others. During UV-induced vitamin D synthesis, numerous substances are produced in the skin—including 20,23-dihydroxyvitamin D_3, 25(OH)T, 20S(OH)T, $20(OH)D_3$, $17,20,23(OH)_3D_3$, $24,25(OH)_2D_3$ (Fig. 1.5)—with physiological effects that are still unknown or little studied (Chaiprasongsuk et al. 2019; Reichrath 2021). They often do not exert their main effect via the classical vitamin D signaling pathway (which is mediated by binding and activation of the VDR), but by binding and activating recently identified alternative receptors (including VDR, AhR, LXR, PPAR, RORs). Recent research suggests that these receptors regulate many important signaling pathways that are relevant for cancer and metabolic diseases (Reichrath 2021). The formation of these receptors in target cells is crucial for the fine-tuning of the vitamin D effect. The tissue-specific production of the different receptor molecules not only makes these tissues targets of the vitamin D effect, but also determines the exact manifestation of these effects. We now know that these receptors regulate several thousand of the approximately 20,000 coding genes in our genome, directly or indirectly, after their activation. Through these mechanisms, the diverse functions of the vitamin D system can be temporally and spatially coordinated to meet the physiological needs of the different organs. In other words, depending on the amount of these enzymes, vitamin D metabolites and corresponding receptors present in a cell, the most diverse effects can be mediated.

It should be noted that vitamin D metabolites can exert their effect locally (autocrine or paracrine effect) or in more distant organs (endocrine effect, after release of vitamin D metabolites into the blood). Above all, $1,25(OH)_2D_3$ exerts its effect via these two different signaling pathways. In the endocrine signaling pathway, $1,25(OH)_2D_3$ formed in the kidney is released into the blood, allowing it to reach distant organs for the regulation of bone and calcium metabolism. In the autocrine/paracrine signaling pathway, on the other hand, $1,25(OH)_2D_3$ is directly formed in numerous organs for a local effect. Both the kidney (endocrine signaling pathway) and numerous other organs (autocrine/paracrine signaling pathway) require 25(OH)D (the storage form of vitamin D in our body) as a starting material (substrate). Since the kidney, unlike other organs, has special proteins (megalin, cubilin) that facilitate the uptake of vitamin D metabolites, the endocrine signaling pathway can fully fulfill its physiological function (including regulation of calcium and bone metabolism) with 25(OH)D blood levels greater than about 10 ng/ml. The autocrine/paracrine signaling pathway, on the other hand, requires higher 25(OH)D blood levels, according to many experts about 20-40 ng/ml, to ensure its physiological tasks (including prevention of infectious, autoimmune, cardiovascular, cancer diseases) in numerous other organs.

▶ **Note** Biologically active vitamin D metabolites mediate significant parts of their effect through the binding and activation of the vitamin D receptor

(VDR) and other nuclear receptors (including aryl hydrocarbon receptor [AhR], liver X receptor [LXR]). This regulates several thousand target genes. The detection of these receptors identifies organs as targets of the vitamin D effect. Various regulatory mechanisms allow an individual fine-tuning of the required hormone effect to adapt to a variety of different requirements. Thus, it can be individually dosed and locally adapted in the different target cells to the respective needs.

5.3 Misconception 3: *All Effects of Vitamin D Formed in the Skin Can Be Compensated by Oral Intake of Vitamin D*

5.3.1 Correction

This statement is incorrect! Because in the UV-induced cutaneous vitamin D synthesis, exclusively various biologically active vitamin D metabolites are formed, which do not occur after oral intake of vitamin D.

5.3.2 Comment

5.3.2.1 Biologically Active Vitamin D Metabolites are Formed in the Skin, Which Do Not Occur After Oral Intake of Vitamin D

For example, in the context of UV-induced vitamin D synthesis in the skin, biologically active tachysterol derivatives are formed from the vitamin D precursor previtamin D (Chaiprasongsuk et al. 2019). In addition, the minor structural differences between the vitamin D_3 produced in the skin and the vitamin D_2 that usually makes up a significant part of the vitamin D ingested with food also cause functional differences, for example, affecting their half-life (Reichrath 2021). This correlation suggests that oral intake of vitamin D does not guarantee all the positive effects of UV-induced vitamin D synthesis in the skin.

▶ **Note** In the UV-induced synthesis of vitamin D, biologically active metabolites are formed in the skin, which are not formed after oral intake of vitamin D.

References

Chaiprasongsuk A, Janjetovic Z, Kim TK, Jarrett SG, D'Orazio JA, Holick MF, Tang EKY, Tuckey RC, Panich U, Li W, Slominski AT. Protective effects of novel derivatives of vitamin D_3 and lumisterol against UVB-induced damage in human keratinocytes involve activation of Nrf2 and p53 defense mechanisms. Redox Biol. 2019;24:101206. https://doi.org/10.1016/j.redox.2019.101206. Epub 2019 Apr 20. PMID: 31039479; PMCID: PMC6488822.

Holick MF, MacLaughlin JA, Clark MB, Holick SA, Potts JT Jr, Anderson RR, Blank IH, Parrish JA, Elias P. Photosynthesis of previtamin D3 in human skin and the physiologic consequences. Science. 1980;210(4466):203–5. https://doi.org/10.1126/science.6251551. PMID: 6251551.

Reichrath J, Herausgeber. Sonne – die Dosis macht's! Berlin/Heidelberg: Springer; 2021.

Wacker M, Holick MF. Sunlight and Vitamin D: a global perspective for health. Dermatoendocrinol. 2013;5(1):51–108. https://doi.org/10.4161/derm.24494. PMID: 24494042; PMCID: PMC3897598.

What is the Optimal Vitamin D Status?

6

Contents

6.1 Misconception 1: *To Assess Vitamin D Status, the Blood Level of the Biologically Active Metabolite 1,25-dihydroxyvitamin D (1,25(OH)$_2$D) Should Be Determined. If this Value is Within the Normal Range, There is No Vitamin D Deficiency* 94
6.2 Misconception 2: *The Determination of the Serum Concentration of the Biologically Inactive Vitamin D Metabolite 25-hydroxyvitamin D (25(OH)D) is Unsuitable for Assessing a Person's Vitamin D Status* 96
6.3 Misconception 3 *An Optimal Vitamin D Status is Assumed When the Blood Level for 25(OH)D is Above 10 ng/ml and there is No Indication of a Disease of the Bone and Calcium Metabolism* 98
References 99

6.1 Misconception 1: *To Assess Vitamin D Status, the Blood Level of the Biologically Active Metabolite 1,25-dihydroxyvitamin D (1,25(OH)$_2$D) Should Be Determined. If this Value is Within the Normal Range, There is No Vitamin D Deficiency*

6.1.1 Correction

Determining the concentration of the classic, biologically active vitamin D metabolite 1,25-dihydroxyvitamin D (1,25(OH)$_2$D) in the blood is not suitable for assessing a person's vitamin D supply for several reasons. Even if this value in the blood is within the normal range and the endocrine effect on bone and calcium metabolism is intact, this does not guarantee optimal vitamin D supply in other organs. A vitamin D deficiency, which is associated with significant and diverse dangers to health, cannot be ruled out in many organs, among other things, for the mediation of autocrine/paracrine effects (see below).

6.1.2 Comment

6.1.2.1 Which Laboratory Parameter is Best Suited to Assess Vitamin D Status?

The determination of the biologically inactive vitamin D metabolite 25-hydroxyvitamin D (25(OH)D) is the most suitable laboratory test for assessing a person's vitamin D status. This blood value indicates "how full the tank is", which not only the kidney, but also all other organs rely on to form the biologically active metabolite 1,25-dihydroxyvitamin D (1,25(OH)$_2$D). The determination of this biologically active vitamin D metabolite in the blood is unsuitable for assessing a person's vitamin D supply. There are several reasons for this. A sufficiently high concentration of 1,25(OH)$_2$D in the blood merely means that the

6.1 Misconception 1: *To Assess Vitamin D Status ...*

kidney produces sufficient amounts of this biologically active metabolite and releases it into the blood for endocrine effects. However, this does not allow conclusions to be drawn about the vitamin D supply in other organs. This is because the kidney is the only organ in our body that has special and very efficient uptake mechanisms for vitamin D (which are based on two proteins referred to as megalin/cubilin). Therefore, even at relatively low 25(OH)D serum concentrations (>10 ng/ml), it can absorb enough 25(OH)D from the blood and produce sufficient $1,25(OH)_2D$. Its release into the blood ensures unrestricted endocrine vitamin D action (which is particularly important for bone health and the regulation of calcium metabolism).

But what about other organs? The biologically active $1,25(OH)_2D$, which is not formed in the kidney but locally in other tissues, is not released into the blood. Instead, it locally controls important autocrine/paracrine signaling pathways in the respective tissues and is important for the prevention and therapy of cancer, infectious, metabolic and autoimmune diseases. Since these organs, unlike the kidney, do not have special uptake mechanisms, they require higher 25(OH)D serum concentrations (>20 ng/ml), according to many experts, to absorb sufficient 25(OH)D from the blood and convert it into the active metabolite $1,25(OH)_2D$.

Another reason why determining the concentration of the biologically active vitamin D metabolite 1,25-dihydroxyvitamin D ($1,25(OH)_2D$) in the blood is unsuitable for assessing a person's vitamin D supply is that this laboratory value, due to its short half-life of about 1–3 days, is subject to large fluctuations and thus only represents a snapshot that does not allow a reliable statement about the longer-term supply situation.

▶ **Note** Determining the concentration of the biologically active vitamin D metabolite 1,25-dihydroxyvitamin D ($1,25(OH)_2D$) in the blood is not suitable for assessing a person's vitamin D supply.

6.2 Misconception 2: *The Determination of the Serum Concentration of the Biologically Inactive Vitamin D Metabolite 25-hydroxyvitamin D (25(OH)D) is Unsuitable for Assessing a Person's Vitamin D Status*

6.2.1 Clarification

The determination of the serum concentration of the biologically inactive vitamin D metabolite 25-hydroxyvitamin D (25(OH)D) is currently considered the most suitable blood test for assessing a person's vitamin D status.

6.2.2 Comment

6.2.2.1 How Full is the Tank?

The biologically inactive vitamin D metabolite 25-hydroxyvitamin D (25(OH)D) is the storage form of vitamin D in our body (Reichrath 2021). Its blood level thus indicates "how full the tank is" that both the kidney and other organs can draw on to form the biologically active metabolite 1,25-dihydroxyvitamin D (see also above).

▶ **Note** The biologically inactive vitamin D metabolite 25-hydroxyvitamin D (25(OH)D) is the storage form of vitamin D in our body. The determination of its blood concentration is currently considered the most suitable test for assessing the vitamin D supply (the vitamin D status) of a person.

How is it possible for the vitamin D hormone system to perform so many different tasks in almost all organs of our body in a timely manner? How can it exert such a wide range of different effects? One reason certainly lies in the complex structure of vitamin D metabolism. This allows it to be regulated and fine-tuned at many different points. These interconnections enable our body to adapt to various physiological and pathophysiological conditions. The fine-tuning is mainly achieved through the formation of different vitamin D metabolites, which is mediated by specific cytochrome P450 enzymes and proceeds in several stages. In addition, the formation of corresponding receptors (including VDR, AhR, LXR, PPAR, RORs) in target cells is crucial for the regulation of vitamin D effects. Both the synthesis (including CYP2R1, CYP27A1, CYP27B1, CYP11A1) and the metabolism (degradation, including CYP24A1) of the numerous vitamin D metabolites are determined by these enzymes. These decide which vitamin D metabolites are present at which time in which quantity and where they exert their effect.

Which of the various corresponding receptor types are present in the respective target cells is also of essential importance for the quality of the vitamin D effect. The binding and activation by vitamin D metabolites makes the differential expression of these different receptors in the individual organs not only targets of the vitamin D effect, but also determines the exact expression of these effects. We know today that these receptors regulate several thousand of the approximately 20,000 coding genes of our genome directly or indirectly after their activation. This allows the diverse tasks and functions of the vitamin D system to be adapted to the respective physiological needs of the different organs in a timely and spatially coordinated manner. In other words: Depending on the quantity of these enzymes, vitamin D metabolites and corresponding receptors present in a cell, the most diverse effects can be mediated and tasks fulfilled. It should also be noted that vitamin D metabolites can exert their effect locally (autocrine or paracrine effect) or in more distant organs (endocrine effect after release of vitamin D metabolites into the blood).

Not all cells are equally capable of absorbing vitamin D metabolites. This phenomenon is partly responsible for the fact that to ensure the various vitamin D effects (endocrine or autocrine/paracrine), different "optimal" 25(OH)D serum concentrations may be required. The kidney in particular plays a special role here. Because according to current knowledge, it is the only one of our numerous organs that has special mechanisms (so-called megalin/cubilin system) that enable it to produce sufficient amounts of biologically active 1,25-dihydroxyvitamin even at relatively low 25(OH)D levels in the blood (greater than about 10 ng/ml) and to release it into the blood. In contrast, the other organs require significantly higher 25(OH)D levels in the blood (greater than about 20 ng/ml) for sufficient synthesis of 1,25-dihydroxyvitamin D for autocrine/paracrine effects.

6.3 Misconception 3: *An Optimal Vitamin D Status is Assumed When the Blood Level for 25(OH)D is Above 10 ng/ml and When there is No Indication of a Disease of the Bone and Calcium Metabolism*

6.3.1 Correction

According to current scientific knowledge, a 25(OH)D blood level above 10 ng/ml is usually sufficient to ensure endocrine functions and prevent the occurrence of diseases of the bone and calcium metabolism. However, to maintain autocrine/paracrine effects, higher concentrations are necessary according to current data, probably values above at least 20 ng/ml are required. This is supported by numerous epidemiological and experimental studies (see also Sect. 6.1.2.1).

6.3.2 Comment

6.3.2.1 What is the Optimal Vitamin D Status?

It can be assumed that with 25(OH)D blood levels between 10 and 20 ng/ml, there is an increased risk for the occurrence and an unfavorable clinical course of numerous cancer, infectious, autoimmune and metabolic diseases. However, there are indications that higher 25(OH)D serum levels may be required for optimal supply. Here are two brief examples: It has been shown that the breast milk of nursing mothers contains sufficient vitamin D to supply the infants when the mothers' 25(OH)D serum level is more than 40 ng/ml. Indigenous peoples, who, for example, live

Fig. 6.1 Do Maasai or other indigenous peoples near the equator have the optimal vitamin D status? Men from the Maasai tribe, who have reported 25(OH)D serum levels of about 46 ng/ml, demonstrate their muscle strength. (*Source*: Wacker and Holick 2013, with kind permission from Dr. Holick)

in the Horn of Africa under conditions that most closely resemble those intended for humans from an evolutionary perspective, show 25(OH)D serum levels of more than 40 ng/ml (Fig. 6.1) (Reichrath 2021; Wacker and Holick 2013).

▶ **Note** With blood levels for 25(OH)D between 10 ng/ml and 20 ng/ml, there is a vitamin D deficiency, even though the bone and calcium metabolism is not affected.

References

Reichrath J, Herausgeber. Sonne – die Dosis macht's! Berlin/Heidelberg: Springer; 2021.
Wacker M, Holick MF. Sunlight and vitamin D: a global perspective for health. Dermatoendocrinol. 2013;5(1):51–108. https://doi.org/10.4161/derm.24494. PMID: 24494042; PMCID: PMC3897598.

Conclusion with Practical Recommendations for Ensuring Good Vitamin D Supply while Safely Interacting with the Sun

Contents

References 109

How do we achieve the goal of optimal vitamin D supply while being health-conscious in dealing with the sun? What does *"health-conscious handling of the sun"* mean in practice? Is there such a *"golden path"* at all? Or do we only have the choice between Scylla and Charybdis, i.e., between vitamin D deficiency and skin cancer risk? To help you clarify these questions, I have designed the step-by-step scheme in Fig. 7.1 with my personal recommendations. My goal is to offer you a concept that is above all easy to implement, but also takes into account individual factors and can be applied at all times of the year.

If the vitamin D status is assumed as an indicator for sufficient use of the sun, more than 60% of the German population have a deficiency of sunlight, even in the summer. In this context, it should also be noted that not all positive effects of sunlight can be replaced by taking vitamin D-containing dietary supplements. Therefore, it is assumed that only a minority of the population uses the sun sufficiently under our living conditions.

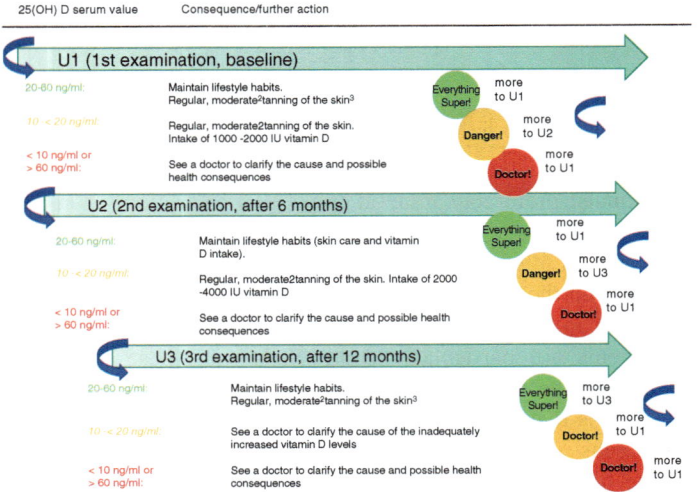

Fig. 7.1 Vitamin D traffic light with recommendations for optimal vitamin D status. From the determination of the 25-hydroxyvitamin D concentration in the blood, different situations arise, which are marked analogously to a traffic light with green, yellow or red

Therefore, in my conviction, the positive health effects outweigh the possible risks for the majority of our population with a cautious, responsible increase in exposure to sunlight ("moderate sunbathing").

The aim of this step-by-step scheme is to fully utilize the positive effects of sunlight with as little health risk as possible through responsible handling of sunlight. More precisely, these recommendations for dealing with the sun ("*Vitamin D/Sun Traffic Light*") aim to avoid skin cancer on the one hand and ensure sufficient supply of vitamin D on the other. The goal is thus to optimize the benefit/risk profile when dealing with sunlight. First, I would like to introduce this step-by-step scheme theoretically and then present its practical implementation.

For this concept, the determination of the vitamin D status is of central importance. This not only allows an assessment of a person's vitamin D supply, but can also serve as an indicator of whether one is sufficiently exposed to sunlight. The step-by-step scheme is based on two pillars: exposure to natural sunlight and the intake (oral supplementation) of vitamin D as a dietary supplement. To assess in what way exposure to natural sunlight and according to what scheme oral supplementation with vitamin D should take place, the 25-hydroxyvitamin D (25(OH)D) level in the serum is first determined. This value is generally accepted as the most suitable laboratory parameter for assessing a person's vitamin D status. With this value, it can therefore be determined how well the vitamin D tank is filled. Until a few years ago, some experts still held the view that it was not advisable to measure the 25(OH)D level in the serum in our population on a large scale. Rather, due to the fact that over 60% of people in Germany have a vitamin D deficiency, a general substitution with vitamin D was recommended. This recommendation was justified by the high costs of determining the vitamin D status, the widespread prevalence of vitamin D deficiency, as well as the low costs and great safety of oral intake of vitamin D preparations. However, in my conviction, it is important to know the individual vitamin D status as a prerequisite for healthy handling of sunlight. Also, the costs for determining the 25(OH)D concentration in the blood have significantly decreased in recent years. Unfortunately, it is still a reality that in many cases the costs for determining the 25(OH)D concentration in the blood are not covered by health insurances and that one has to bear these costs oneself.

The practical implementation of the step-by-step scheme described here relies not only on the determination of vitamin D status but also on information from the scientific literature. This includes estimates of the effects of vitamin D formation in the skin and its administration as a dietary supplement on vitamin D status. It has been published that unprotected, moderate sun exposure of about 25% of the body surface (e.g., hands, arms, and face) in a dose of 25% of the MED (minimal erythema dose; lowest radiation dose at which the skin begins to redden) is roughly equivalent to the oral intake of 800–1500 IU of vitamin D. To achieve a sufficient vitamin D status, it would therefore often be sufficient to expose oneself to the sun 4 to 6 times a week from April to September. This corresponds to a stay in the sun of about 10–20 minutes each time for the prevailing skin type II–III in Europe, depending on other factors such as time of day and season. According to the results of well-designed studies (Cashman et al. 2008, 2009; Cashman 2020), a vitamin D intake (as a dietary supplement/supplement) of 1000 IU (25 µg) per day achieves a sufficient vitamin D status (25(OH)D blood concentration >20 ng/ml; >50 nmol/l) in about 95–97.5% of the Irish population during the winter months. American endocrinologists and many other experts consider a daily intake of up to 10,000 IU of vitamin D in adults (>18 years) to be safe, which corresponds to a total amount of about 1,800,000 IU of vitamin D in 6 months.

In my concept, the vitamin D status is regularly determined at 6-month intervals. Depending on the examination results, there are several options for further action. If the initial value (1st examination) of the 25(OH)D serum concentration is between 20 and 60 ng/ml, everything is "super", and lifestyle habits, including sun exposure, should then be maintained unchanged. A re-examination should then take place after 6 months (also to take into account seasonal influences). If the 25(OH)D initial value is between 10 and 20 ng/ml, there is already a health-threatening vitamin D deficiency. Then the skin should be exposed to the sun regularly and moderately, and additionally, 1000–2000 IU of vitamin D should be taken daily. In this case, too, a repeat blood test is advisable after 6 months. If the 25(OH)D initial value is <10 or >60 ng/ml, there may be a serious health-threatening disorder of vitamin D metabolism. In this case, a doctor should be consulted for further clarification.

After 6 months, the 2nd examination (U2) follows. Regardless of the initial situation, if the 25(OH)D blood values are between 20 and 60 ng/ml, everything is again "super". The most recently practiced lifestyle habits (sun exposure and oral intake of vitamin D) should in this case be maintained unchanged, and a re-examination should take place after 6 months. If the 25(OH)D blood value at this examination time is between 10 and 20 ng/ml, there is a health-threatening vitamin

D deficiency. Depending on the intake of vitamin D practiced at this time, various approaches can be derived in addition to the continuation of regular and moderate sunbathing of the skin. If no vitamin D supplements have been used so far, the daily intake of 1000–2000 IU of vitamin D should now be started. If 1000–2000 IU of vitamin D are already being taken, this dose should be increased to 2000–4000 IU daily. A repeat blood test should then take place after another 6 months (then re-evaluation as at U3). If the 25(OH)D blood values are <10 or >60 ng/ml at U2, there may be a serious health-threatening disorder of vitamin D metabolism (for the safety of vitamin D, see also Sect. 1.2.2.9 and Chap. 3). In this case, a doctor should be consulted for further clarification.

After 12 months, the 3rd examination (U3) follows. If the 25(OH)D blood levels are then between 20 and 60 ng/ml, everything is again "super". The last practiced lifestyle habits (sun exposure and oral intake of vitamin D) should be maintained unchanged. Subsequently, another blood test should be carried out after a further 6 months (assessment again as at U3). If the 25(OH)D blood level at this examination time (U3) is between 10 and 20 ng/ml, there is a health-threatening vitamin D deficiency. Depending on the intake of vitamin D practiced at this time, different approaches can be derived. If only 1000–2000 IU of vitamin D are taken daily, this dose should be increased to 2000–4000 IU. Another blood test should then be carried out after a further 6 months (re-assessment again as at U3). However, if these too low 25(OH)D blood levels occur while taking 2000–4000 IU of vitamin D, it is already necessary to consult a doctor to clarify the cause and possible consequences. If the 25(OH)D blood levels at the examination time U3 are <10 or >60 ng/ml, there may be a serious health-threatening disorder of vitamin D metabolism. In this case, a doctor should be consulted for further clarification.

The vitamin D traffic light is a guide for healthy adults (>18 years) without special dietary habits ("Standard Western Diet"), without additional exposure to artificial UV radiation (e.g. solarium, dermatological phototherapy) and without individual reasons that oppose exposure to natural sunlight (e.g. photosensitizing drugs, increased light sensitivity of the skin). The effects of certain factors such as location, season are taken into account in this approach. Although under certain conditions (e.g. location north of the 50th latitude, October to March) no significant vitamin D production in the skin is possible due to the low UV-B component of sunlight, the skin should still be exposed to the sun unprotected under these conditions to allow non-UV-dependent positive effects of sunlight (e.g. formation of "happiness hormones", melatonin). It can be assumed that this will achieve a sufficient vitamin D status.

U1: 1st Examination (Baseline)
U1—"Green Light"

Everything's great, keep it up!

Vitamin D status in target range *(25-hydroxyvitamin D (25(OH)D) levels in blood 20–60 ng/ml)*.

Recommendation: Maintain lifestyle habits, you're doing everything right. This way of dealing with the sun produces enough vitamin D in the skin without increasing the risk of melanoma (expose about 25–50% of the body surface to the sun daily with up to 50% of the MED if possible). Repeat U1 after 6 months.

U1—"Red Light"

Warning, see a doctor!

Severe vitamin D deficiency *(25-hydroxyvitamin D concentration in blood < 10 ng/ml)* or too much vitamin D *(25-hydroxyvitamin D concentration in blood >60 ng/ml)*.

Recommendation: The vitamin D status is a serious health risk. Therefore, see a doctor to clarify the cause and health consequences. Then repeat U1.

U1—"Yellow Light"

Caution, change lifestyle habits!

Moderate vitamin D deficiency *(25-hydroxyvitamin D concentration in blood 10-< 20 ng/ml)*.

Recommendation: The vitamin D status is too low and therefore a health risk. The skin should be exposed to the sun consequently (daily about 25–50% of the body surface with up to 50%

of the MED); additionally take 1000–2000 IU of vitamin D daily as a supplement. Continue with U2 after 6 months.

U2 (2nd Examination after 6 months)
Control of vitamin D status after implementing the approach recommended in the 1st examination for about 6 months (also to take into account seasonal effects).

Because from October to March, natural sunlight at a location north of the 50th latitude does not cause significant synthesis of vitamin D in the skin. An oral intake of 1000–2000 IU of vitamin D daily corresponds to an amount of about 180,000–360,000 IU of vitamin D in 6 months. The resulting *minimum total amount* from UV exposure and oral intake is therefore about 180,000 IU of vitamin D in 5 months according to this approach. According to the results of well-designed studies (Cashman et al. 2008), in the Irish population during the winter months, a vitamin D intake (as a supplement) of 1000 IU (25 μg) per day achieves a sufficient vitamin D status (25-hydroxyvitamin-D serum concentration >20 ng/ml; >50 nmol/l) in about 95–97.5% of the population. According to literature, from March to September (at a location north of the 50th latitude), moderate, natural sun exposure (about 25–50% of the body surface with up to 50% of the MED) can cause the synthesis of up to 3000 IU of vitamin D daily in the skin. This corresponds to an amount of up to about 540,000 IU of vitamin D in 6 months. The additional oral intake of up to 2000 IU of vitamin D daily adds up to about 360,000 IU of vitamin D in 6 months. The resulting *maximum total amount* from UV exposure and oral intake is therefore about 900,000 IU of vitamin D after 6 months according to this approach.

U2—"Green Light"
Everything is great, keep it up!

Vitamin D status in the target range (*25-hydroxyvitamin-D level in the blood 30–60 ng/ml*).

Recommendation: You are currently doing everything right, lifestyle habits should be maintained. Through this interaction with the sun, the skin produces sufficient vitamin D without increasing the risk of melanoma (about 25–50% of the body surface if possible daily exposure to up to 50% of the MED of the sun). Continue to take 1000–2000 IU of vitamin D daily as a dietary supplement. Continue with U1 after 6 months.

U2—"Red Light"

Attention, danger, see a doctor!

Severe vitamin D deficiency *(25-hydroxyvitamin-D concentration in the blood <10 ng/ml)* or over-supply of vitamin D *(25-hydroxyvitamin-D concentration in the blood >60 ng/ml)*.

Recommendation: The vitamin D status is a serious health risk. It is essential to see a doctor to clarify the cause and health consequences. Then continue with U1.

U2—"Yellow Light"

Caution, change lifestyle habits!

Moderate vitamin D deficiency *(25-hydroxyvitamin-D concentration in the blood 10-< 20 ng/ml)!*

Recommendation: The vitamin D level is too low, therefore change lifestyle habits. Increased sun exposure (daily up to about 25–50% of the body surface with up to 50% of the MED); additionally increase the daily vitamin D intake as a dietary supplement to 2000–4000 IU. Depending on the season and other factors, the total supply of vitamin D in these 6 months corresponds to the oral intake of at least 360,000 (oral intake of 2000 IU of vitamin D daily over 6 months, no cutaneous vitamin D production)—1,260,000 IU (oral intake of 4000 IU daily, cutaneous production of 3000 IU daily, each over 6 months) of vitamin D. American endocrinologists and many other experts consider a daily intake of up to 10,000 IU of vitamin D in adults (>18 years) as safe, which corresponds to a total amount of about 1,800,000 IU of vitamin D in 6 months. Continue with U3 after 6 months.

U3: (3rd Examination after 12 Months)

Recheck of the vitamin D status, after the approach recommended at the 2nd examination has been carried out for about 6 months.

U3—"Green Light"

Everything is great, keep it up!

Vitamin D status in the target range *(25-hydroxyvitamin D levels in the blood 20–60 ng/ml)*.

Recommendation: You are doing everything right, maintain lifestyle habits. The skin should continue to be exposed to sunlight in such a way that sufficient production of vitamin D is ensured without an increased risk of melanoma (daily approx. 25–50% of the body surface with up to 50% of the MED). In addition, continue to take 2000–4000 IU of vitamin D daily as a dietary supplement. After 6 months, continue with U3.

U3—"Red Light"

Attention danger, see a doctor!

Severe vitamin D deficiency *(25-hydroxyvitamin D concentration in the blood <10 ng/ml)* or too much vitamin D *(25-hydroxyvitamin D concentration in the blood >60 ng/ml)*.

Recommendation: The vitamin D status is not okay and is a serious health risk. It is essential to see a doctor to clarify the cause and consequences. Then continue with U1.

U3—"Yellow Light"

Caution, see a doctor!

Moderate vitamin D deficiency *(25-hydroxyvitamin D concentration in the blood 10–< 20 ng/ml)*.

Recommendation: The vitamin D level is too low. This is very unusual if the last recommendations have been correctly implemented. Therefore, see a doctor to clarify the cause and consequences. Then continue with U1.

References

Cashman KD. Vitamin D deficiency: defining, prevalence, causes, and strategies of addressing. Calcif Tissue Int. 2020;106(1):14–29. https://doi.org/10.1007/s00223-019-00559-4. Epub 2019 May 8. PMID: 31069443.

Cashman KD, Hill TR, Lucey AJ, Taylor N, Seamans KM, Muldowney S, Fitzgerald AP, Flynn A, Barnes MS, Horigan G, Bonham MP, Duffy EM, Strain JJ, Wallace JM, Kiely M. Estimation of the dietary requirement for vitamin D in healthy adults. Am J Clin Nutr. 2008;88(6):1535–42. https://doi.org/10.3945/ajcn.2008.26594. PMID: 19064513.

Cashman KD, Wallace JM, Horigan G, Hill TR, Barnes MS, Lucey AJ, Bonham MP, Taylor N, Duffy EM, Seamans K et al. Estimation of the dietary requirement for vitamin D in free-living adults >=64 y of age. *Am. J. Clin. Nutr.* 2009;89:1366–1374. https://doi.org/10.3945/ajcn.2008.27334.

GPSR Compliance

The European Union's (EU) General Product Safety Regulation (GPSR) is a set of rules that requires consumer products to be safe and our obligations to ensure this.

If you have any concerns about our products, you can contact us on ProductSafety@springernature.com

In case Publisher is established outside the EU, the EU authorized representative is:

Springer Nature Customer Service Center GmbH
Europaplatz 3
69115 Heidelberg, Germany

Batch number: 09750407

Printed by Printforce, the Netherlands